# Technology 2001

# Technology 2001
## The Future of Computing and Communications

*edited by Derek Leebaert*

The MIT Press
Cambridge, Massachusetts
London, England

Second printing, 1991

© 1991 Massachusetts Institute of Technology

This book was set in Baskerville by Compset, Inc. and was printed and bound in the United States of America.

Library of Congress Cataloging-in-Publication Data

Technology 2001 : the future of computing and communications / edited
  by Derek Leebaert.
    p.   cm.
  Includes index.
  ISBN 0-262-12150-6
  1. Computers and civilization. 2. Telecommunication.
I. Leebaert, Derek. II. Title: Technology two thousand one.
III. Title: Technology two thousand and one.
QA76.9.C66T34   1991
303.48′34—dc20                                               90-40022
                                                                CIP

for Jane White, who sees the future

# Contents

# 12

# Foreword

It is more than fifty years since I first gazed entranced at a wonderful mass of gleaming brass gears and levers enshrined in a glass case in the Science Museum at South Kensington. I do not know why the unfinished fragment of the Babbage computer fascinated me, or why I visited it so often in those distant days of my leisurely prewar Civil Service career.

Little did I imagine that, some twenty years later—after becoming a modest appendage of MIT's Radiation Lab—I should be examining the fragments of a much earlier machine, built by a Greek genius whose name we shall never know. In 1965 I persuaded the curators of the Athens Museum to show me the Antikythera Mechanism, then ignominiously relegated to a cigar box in the basement.

It is sobering to realize how little progress was made in the 2000 years between Antikythera and Charles Babbage—and how little progress would have been made by now had it not been for the advent of electronics. Barring historical accidents, Newton might have possessed a pretty good mechanical desk calculator—and how *that* would have speeded up his work! But it would have been a technological dead end—a blind alley down which no further progress could have been made.

For me, the era of the electronic calculator began at a NASA conference in 1970 when Hewlett-Packard's Bernard Oliver displayed the prototype HP35 "Electronic Slide Rule." (Was this the last major example of American technological pioneering, before inspiration departed to the Land of the Rising Sony?) Since then, developments in the field of computers have been so swift that yesterday's miracle is today's obsolescent junk.

As I leaf in baffled incomprehension through the current issue of *Byte,* I sometimes try to imagine what a scientist of the late paleoelectronic era (circa 1960) would have thought of its contents. I've certainly seen ample proof in my own lifetime of Clarke's Third Law: "Any sufficiently advanced technology is indistinguishable from magic." Not in my wildest dreams did I ever imagine that within a decade slide rules and mathematical tables would be obsolete—or that I would have the equivalent of several million vacuum tubes sitting on my office desk. (Soon they'll be in the palm of my hand, and a little later they may be floating around in my bloodstream.)

Since Derek Leebaert has chosen to insert a certain notorious date into this book's title, I suppose I should answer the inevitable question: "Do you really think HAL will arrive on schedule?" My answer is No—but I'm damn sure he'll be here by 2100, which is only tomorrow in human history, and a millisecond away on the cosmic time scale (which should now be our increasing concern).

I am well aware that many profound thinkers—among them Roger Penrose, in *The Emperor's New Mind*—doubt that computers will ever achieve real intelligence or self-awareness (whatever those terms mean). I find this hard to believe, as I can see no reason why there should be any fundamental difference between a device constructed of silicon and one of carbon. The distinction may turn out to be just as false as that once drawn, with equal confidence, between inorganic and organic chemistry.

Almost thirty years ago I pondered the future of computers in "The Obsolescence of Man," the penultimate essay in *Profiles of the Future*. The conclusions I arrived at then have, if anything, been reinforced by later developments and the speculations of better-qualified authors. (See Hans Moravec's *Mind Children,* perhaps the most stunning extrapolation of this theme.)

Here are the views I expressed in 1961; I doubt that they will be different in 2001:

The popular idea, fostered by comic strips and the cheaper forms of science fiction, that intelligent machines must be malevolent entities hostile to man, is so absurd that it is hardly worth wasting energy to refute it. I am almost tempted to argue that only unintelligent machines can be malevolent; anyone who has tried to start a balky outboard will probably agree. Those who picture machines as active enemies are merely projecting their own aggressive instincts, inherited from the jungle, into a world where such things do not exist. The higher the intelligence, the greater the degree of cooperativeness. If there is ever a war between men and machines, it is easy to guess who will start it.

Yet however friendly and helpful the machines of the future may be, most people will feel that it is a rather bleak prospect for humanity if it ends up as a pampered specimen in some biological museum—even if that museum is the whole planet Earth. This, however, is an attitude I find impossible to share.

No individual exists for ever; why should we expect our species to be immortal? Man, said Nietzsche, is a rope stretched between the animal and superman—a rope across the abyss. That will be a noble purpose to have served.

*Arthur C. Clarke*

# Acknowledgments

Two institutions have been pivotal to the writing of *Technology 2001:* the Computer and Business Equipment Manufacturers Association (CBEMA), which is the research and public policy center for the leading manufacturers that collectively generate 4.8 percent of the United States' gross national product, and Future Technology, Inc., an information engineering and technology consulting firm, based in Washington, that provides computer systems solutions for government and industry.

At CBEMA, John Pickitt and Maryann Karinch helped assure unique access to the world's most innovative companies. The CEOs, other senior executives, and chief scientists of these multinational enterprises greatly facilitated this wide-ranging research effort. At Future Tech, my partners—Wallace Seward, Rob Stewart, Dimi Meader, Jeff Howard, Shelby Tucker, Mike Peterson, Nabig Khoury, and John Freck—patiently instructed this economist in microelectronics; Stephanie Seward was perhaps even more patient as she offered technical and administrative support, and Amy Stephenson filled a vital role as special assistant and special friend.

One particular teaching colleague and Future Tech advisor—Timothy Dickinson—was indispensably powerful in bringing together science, mathematics, history, and, most of all, friendship in charting the information age's coming decades. No writer could have a better critic, co-author, and friend.

In Cambridge, Frank Urbanowski, Director of The MIT Press, has shown me the intricacies of publishing for more than fifteen years. And Frank Satlow, the Press' Executive Editor, led this volume to daylight, as did manuscript editor Paul Bethge. I remain grateful to Paul Doty, Mallinkrodt Professor of Biochemistry Emeritus at Harvard, for first appointing me to the Center for Science and International Affairs, which he founded, and to Ray Vernon, Dillion Professor of International Business Emeritus, for asking me to join him in creating and editing the *Journal of Policy Analysis and Management.*

Anne Day Thacher, currently at Harvard University's Dumbarton Oaks, provided constant enthusiasm in Cambridge and in Washington, as did Columbia University's Anne McCormick, who also illuminated the complexities of statistical analysis in the high-technology sector.

The guides within industry and finance who shaped my thoughts and interpretations during this work include Bruce Marquardt, Coopers & Lybrand's Phil Odeen, Price Waterhouse's Rick Cooper, Bob Dean of Ball Aerospace, Manufacturers Hanover's John Price, Advisory Board president David G. Bradley, Abt Associates' Robert Coulam, SoundView Financial CEO Peter C. Wright, Charlotte LeGates, Yale School of Organization and Management professor Paul Bracken, Johns Hopkins professor Michael Mandelbaum, and, from the venture capital community, Richard H. Missner and Maria deGraaf.

From Washington, encouragement as well as insights into America's technological future were offered by Marc Wall and David Webster at the State Department, Colonel Murph McCloy, USMC, the US Treasury's John Hauge, the Defense Department's Bruce Weinrod, and Al Gore.

I'm also grateful to Dr. George Zimmer of the Hungarian Academy of Sciences' Microelectronics Research Institute for his partnership.

Special thanks are owed to Ed Stephenson, and, certainly, to my father, Onno Leebaert.

*Derek Leebaert*
Washington, D.C.

# Technology 2001

# Later Than We Think: How the Future Has Arrived

*Derek Leebaert*

"Any sufficiently advanced technology is indistinguishable from magic," as Arthur C. Clarke, technological visionary and author of science fiction, observes in the foreword. This book sets out to penetrate the magic. It anticipates the achievements of computer technology while speculating about what will have been accomplished by the opening of the new millennium, in 2001 (the year to which Clarke introduced us in *2001: A Space Odyssey*).

*Technology 2001* is the first book in which American scientists and strategic planners writing from inside the computer and communications industry explore the future of the information age across the entire wavefront. In 1963, another group of writer-scientists were marshaled into authoring a science-fiction compendium as "expert dreamers."

This year's science is increasingly difficult to distinguish from last year's sheer romance. With new orders of computational power swelling like thunderheads on the clearly visible horizon, and with the structures of life uncurling in our hands, we are living less the prophecies than the dreams of a generation. In retrospect, our birth years seem like the times of costume dramas. Our present is surely like science fiction relative to the world of our childhood.

Although the individual authors focus on their areas of special expertise, this book is a story of cooperation, as is the information era itself. The bigger we know the world to be, the more we have to reach out into "the world's debate" (to borrow a phrase from Shakespeare).

Implicit in all discussions of technology is a reckoning of society. To declare something technical is to declare it predictable. Instruments shape the way we think, by which implicitly we mean our modes of reasoning, imagining, and remembering.

Imagination shifts with capacity. The toil of genius that won the basics of ancient astronomy now lies in a high-school student's laptop computer. The dissolution of the time costs of distance and the overcoming of the fatigue of repetition accelerate the interplay between technology and human practices. The authors therefore discuss the great structures of knowledge—its organization, access, and employment—over which new orders of scientific modeling, management, communications, finance, marketing, policing, and education—just for starters—will develop exponentially.

One of the book's objectives is to remind institutions how dangerously behind the times they are—to offer a necessary glimpse of the unknowns that are piling up on us. Exclusionary lines are quickly being erased from organization charts. The differential advantage accorded to specific ability is annulled by a lack of awareness of what the new technologies offer. A talented person unable to mobilize computer and communication skills is like a 17th-century mathematician who let calculus pass him by. No one can be assumed to be inaccessible or irrelevant. No natural "capital"—no university or city—dominates innovation; it can happen anywhere. There is a cosmopolitan assumption of capacity. A hundred and fifty years ago, Bolyai and Lobachevski unraveled non-Euclidean geometry on the distant boundaries of Europe, to the astonishment of Western mathematicians. Now, Pakistani computer-store owners can hatch a virus that bedevils two continents.

Technology dissolves particularity as we move ever closer to real-time interaction. Newton might wait a month to get a response to one of his hypotheses. Argument is now conducted almost the same way whether it is across the planet or in one room. In the past, a few universities (such as Cambridge) or a few families (such as the Bernouillis) were preeminent in mathematics. Now anyone who sets to work can be carried on the great currents of the world. For several centuries we have been moving away from the assumption that the test of ability is knowledge itself; understanding is, more and more, about processes rather than data. Will it be considered even less necessary in the future to retain information when it can be looked up—and when it will then be fresher and more closely analyzed than if one had studied it in school?

Dynamic information systems are undercutting by the hour any notion that institutions need a centralized, heroic decision-maker, as opposed to enforcers of critical honesty. This degree of access to information makes it certain that there are no "faraway countries about which we know little." People everywhere are contributing. Each has the capacity to mobilize underlying patterns and forces. But the problem is one of unused human capacity, rather than unused computational capacity. The book examines many of the different technological resources as well as the exponentially greater uses that may be made by bringing them together. Yet we don't presume to include every technological innovation; they couldn't all be imagined.

The world can be divided into two camps: the computer-aware individuals and institutions and the computer-inflexible ones. The book discusses the tools and methods that can help society to maximize its employment of these technologies. But we face a paradox. On one hand, everything seems temporary, because we suspect that nearly anything can be improved upon. On the other hand, the externalities—such as pollution and poor education—do not allow the same geometric improvements as, say, the computer analysis of economic statistics. The computer enables us to do things very fast, but everything else is slower and less accessible than the computer.

The authors sometimes disagree with one another. Harry Tennant and George Heilmeier, for example, offer slightly different estimates about the transistor densities and the clock rates that will have been achieved by 2001 than Gerhard Parker, Albert Yu, and their colleagues. Lee Hoevel's work on the human interface predicts accomplishments that other authors anticipate from different approaches, such as graphics, dynapaper, and networks. And nearly all the contributors disagree on the precise time required to attain a given objective. Such diversity is unsurprising. What we really want is for the reader to draw lines from all these chapters to his own life and to the changes in himself and in his work.

This introductory essay addresses the business and social as well as the economic changes which the concentration of complex and interrelated technologies bring about. Besides asking what the computer will make possible, it spotlights the intense ambiguities to which the computer will give rise—ambiguities that will be resolved only through experience. It examines the likely curves of development through the 1990s, the consequences for management, and the assumptions about the

United States' competitiveness that underlie so many discussions of the technological future.

Several themes run through the book: the challenge to specializations and organizational partitions; the ever-increasing value of individual creativity; the fact that more and more activities must be resolved into their component bundles of possibilities, thus offering extraordinary challenges to the establishing of context and sequence; and, ultimately, the fact that the whole book is an exploration of degrees of freedom.

The authors are reporting to consumers ten years early. These consumers include specialists who want to see what their fellow specialists are doing, executives who are already heavily dependent on computer services and who must anticipate what they will seek to harness next, and concerned generalists who understand that these technologies are representative of the great phase-change of consciousness into replicable symbolizing intelligence. Whether the reader likes it or not, these technologies are taking him into a future that was inconceivable when he was born.

### The Furnace of the New

Everything is being melted in the furnace of the new. Computer and communication technologies are transforming themselves while penetrating the whole structure of human enterprise. Changes in capital formation, biological research, traffic management, insurance, and countless other horizons of initiative—many of them only now illuminated for the first time by these new powers—are happening at faster-than-biological speed. Henry Ford took 20 years to develop the Model A. No one could tolerate such an industrial pace today. A working lifetime of 42 years is a thousandth of the time from the neolithic period to the paleolithic, a hundredth of the time from the galley to the galleon, and the entire span of the period of practical computers.

Perhaps the only thing analogous to such institution-driven accelerations of creativity is the pressure exerted by war to make two or three generations of technology blossom in a few years. Yet information technologies are now the strategic core of business. It is inconceivable that a company can stay competitive without a first-rate information operation—and yet this would have seemed a gross exaggeration a dozen years ago.

Since knowledge is power, one of the most important functions for information systems will be to distribute that power.

Along the way, the importance of programming and of new strategic applications is escalating rapidly. A decade ago, the most expensive computing function was processing power, followed by memory, followed by data, followed by the network, followed by applications programming. In the 1990s, the most expensive and critical functions are the program (or application), followed by distributed networking, followed by data (still somewhere in the middle, although now we focus on storage), with equivalent memory and processing power having become commodities.

What other sector has, in less than twenty years, inverted its entire set of optimization criteria, simultaneously becoming indispensable to almost everything? Some of today's desktop systems are more powerful than Mission Control's roomful of IBM mainframes in July 1969. We are quickly moving down many tracks that we couldn't afford to travel before.

Ever since it was first painfully worked out, thousands of years ago, systematized information was always assumed to be scarce. The computer reversed this as it became the great non-zero-sum technology. Moving from information scarcity to information hyperabundance means moving into an entirely different world of expectations. Long-term consequences don't yield to radical simplification.

The 1960s was the decade of mathematical application—number crunching, accounting, and so on. The 1970s brought word processing; the 1980s brought desktop publishing and computer-aided drafting and design. The 1990s are seeing the collection, for virtually immediate access, of vast amounts of data for business, government, and all analytical purposes. But in 2001 and beyond, the emphasis will be on far more human functions.

### The Leading Edge

The book's first part amplifies the distant thunder of the new technologies—including supercomputing and microchip creation as well as the three increasingly intertwined classes of computing. Trends in microelectronics are driving concurrent trends in the evolution of computer hardware—displays, printers, plotters, and far more exotic accessories. The price of executing a computer instruction is falling steadily, and the cost of executing an instruction varies greatly from mainframes to minicomputers (middle-range, multi-user systems) to micro-computers (single-user, general-purpose workstations, epitomized by the PC). Minicomputer MIPS (millions of instructions per second) cost less than half as much as mainframe MIPS, and microcomputer MIPS cost about a tenth as much as mini-computer MIPS. The cost of information storage continues to plummet.

The computing environment of 2001 will depend largely on whether the mainframe continues to act solely as "the central processor" for a vast amount of unintelligent terminals or whether it becomes "the host" in collecting ever more information and redistributing it for processing to very intelligent workstations (powerful computers in their own right) over extensive, extremely fast networks. The challenge of the 1990s is to manage multi-sensory information—to hybridize sound, picture, text, and perhaps (down the road) more exotic inputs such as locative sense or smell—and to shove users along the ever-harder-to-establish critical path. For the year 2001, our authors project various autonomous technical possibilities that resemble not merely humanity's logical powers but even its intuitive powers.

The mainframe can still supply unrivaled processing and storage power, as well as redundancy. This costs money: the more power, the more paths, the more circuits, the more silicon, and the more dollars. Thus, we will see systems with tens of thousands of parallel accesses to optical disks, able to share resources. Instead of having a system with a billion bytes on your desk, you will be entering a system with immediate access to 300 billion bytes that are being shared to work on very large objectives. There is no need to put such resources on each desktop when queuing and access theory permit efficient use at one remove.

Memory will cost a few dollars per megabyte by the mid-1990s. At $2 a megabyte, a billion bytes will be available for $2000, and a hundred billion bytes—a thousand Britannicas' worth of information—for the cost of a ranch house in Minneapolis. Economically, mainframes are not well positioned to compete, solely with raw MIPS and megaflops, on the specialized problems that we now tackle with minis, super-minis, and workstations. But, as Denos Gazis expounds in his chapter, mainframes will still be indispensable in long-term roles.

Another concern for corporate mainframe computing is the physical protection of the resource. It will take a few more virus attacks before security is regarded as seriously as it must be. In 1990, there was already an estimated $3 billion market for computer-security products and services. Security is an intractable problem because we walk a continuous tightrope between convenient access, ready exchange of information, and protection. The only feasible solution might be building a computer to which no one can ever gain access. "If the system was perfect," said Joseph Schumpeter (in discussing economics), "it wouldn't work."

Supercomputers, definable only as the fastest available in their generation, have become an international symbol of superior manufacturing and scientific prowess as well as of military capability. They are essential for constructing state-of-the-art microelectronic components and aircraft, as well as for fabricating advanced materials, medicines, and chemicals. They are crucial to the security strategy of the United States.

The race for the fastest machines has become an emblem and a test of the policies that shape the future of US military power as well as that of US science and industry. Kathleen Bernard weighs the role of government in securing these strategic heights. But do we have an appropriate set of magnitudes and interconnections for government-size programs in computing? The space program, after all, involved relatively shallow innovation; it was a matter of consolidating and accelerating established processes rather than making radical breakthroughs. But now we are talking about entering unknown territory. What other opportunities should be reviewed before government backs specific technologies or companies as "chosen instruments"?

Supercomputers let us create a laboratory in a computer, as contrasted with sticking a computer in a laboratory. Computers not only process experimental data but can be excellent places in which to ascertain them. To an astrophysicist, such literally analogical computing might serve as an observatory exploring distant galaxies for billions of years as he draws out the family of predictions that a theory holds hidden within its mathematics. As faster computers attack new sets of problems, how should we regard knowledge that has been thus elicited? Irrespective of government policy or academic debates over evidence, the increase in computing power will be immense. The takeoff is previewed (and diagrammed) in the chapter by Gel-singer, Gargini, Parker, and Yu.

### The Technology of the Third Millennium and the Organization of a Third Industrial Revolution

The book's second part, "Wealth and Mastery," addresses the constant need to find the critical path through information abundance. Max Beerbohm once wrote about Jowett, the master of Balliol, looking up at Dante Gabriel Rossetti painting the Arthurian legend on the rotunda of the Oxford Union and calling to him: "And what were they going to do with the grail when they *found* it, Mr. Rossetti?" How do we think we would apply the increased informational power as we pursued our own transcendent quest?

The computer is no longer simply prized as an important giver of answers. Each great conceptualizing improvement has expanded and transformed the nature of the questions that can be asked, much as calculus and the telescope transformed astronomy. New instruments make new questions—and new wonders—practical. There is no fixed universe of questions. Imaginatively used, the computer enlarges the user's capacities, over and above merely storing the fruits of these capacities.

Tennant and Heilmeier confront the unsettling question raised by the abundance of information. Yet abundance keeps changing its form. The 1990s are simply another stage of abundance, albeit the greatest in history. The inertias and densities of abundance can be met by exercises rooted in past epochs of increased information. Over many centuries, societies have generated simplifying mechanisms to cope with similar challenges. Such devices as the school, the newspaper, and the encyclopedia have been created along a continually unfolding spiral of intellectual development. But dynamic interrogation—the ability to enquire systematically and to bring structure to data sought from the passive body of information—was impossible right up until the time of the punch card. Even then, the constraint imposed by the number of holes was closer to that of handwritten text than to that of the telephone switchboard.

Humanity recognized at different stages that there were new wealths of knowledge which could fruitfully be addressed only with new techniques. When the library at Alexandria contained 700,000 volumes, there was information abundance. Abundance is a natural consequence of progress, but it can be a Greek gift indeed, becoming a glut when improperly put to use. Information becomes real knowledge, as distinct from raw data, only when it and the instruments used to interrogate it are organized in congruent ways.

The most powerful of tools is diminished if the follow-up knowledge is left purely random—if the steering wheel is not in skilled hands. Information itself can be valuable; the mere sight of a train heading toward us makes us jump off the track. But the value of information does not increase with use—only with *alert* use. Mere knowledge renders us less adaptive: the more choices we have, the likelier we are to freeze (like the dog stranded between two bones in medieval philosophy).

Language not only transfers knowledge, it also imparts consciousness and coherence. It brings detailed knowledge into being and contextualizes it. Irrational perspectives are transformed into self-knowledge, and analogies are organized into taxonomies. Language is also a crucial instrument of self-discovery and self-organization. What did I hear? What did I see? Raw unexamined sensory data are basic but not self-explanatory. Acute personal sensation can be perilous: rabies, after all, is heightened sensitivity.

Language is exogenous. Nature created it. It evolved at the lowest level of conscious undertaking as a very specific physical repertoire—the organization of certain actions of the lungs, mouth, and throat. As a complex function that we perform unconsciously, language is an increasingly appropriate analogy for computing. Computing pushes us toward a repertoire that is for the most part unconscious, much as we hunt for a musical theme in random noise while we read, drive, or work at our PC. But to what extent will we find ourselves developing secondary and tertiary sensitivities beyond even the pilot's to the airplane or the hunter's for the jungle? Will such sensitivities be closer to the whale's vast-ranging sensory interaction with the ocean?

Writing, or symbolized knowledge, became the most radical achievement of the consciously acting human mind. Sounds could now be frozen, and transactions too complex to perform in memory could be executed laterally and sequentially in the external world. Events could be experienced outside the circle, or at a later date. Writing is simultaneously the great instrument of order and the first great impersonal stimulus to the intellectual imagination. We can organize and recognize complexities outside ourselves, where memory and imagination are unlikely to be adequate. Writing concentrates our consciousness and raises the value of reflective cognition relative to the value of memory. Writing transcends the physical constraints of distance and duration as well as the imaginative constraints (such as attention span).

Civilization is about the externalization and cross-breeding of knowledge—the tapping of other people's skills. Just as genetic information passes across not just species and genera but also across phyla and kingdoms, human data can combine into the most unanticipated forms. The market is a powerful model of unanticipated information, able to function more accurately through price than through fiat. There is an ongoing transformation of the nature of information, as well as a re-envisioning of the multiplicity of its uses as it soaks into human arrangements along new lines of opportunity.

Once a given body of knowledge is seen as superbly organized, it becomes less relevant. The more informally organized the body of knowledge, the more things can be teased out of it. The organization of knowledge is never perfect in any field that is not formally rigorous. The more lateralization, the more imparted ambiguities. Judgment is summoned up where understanding is incomplete.

Complex questions about systems, biological or social, elicit other modes of judgment. The writing of history, with its wealth of assumptions, is the classic example. The more information we have, the more variables become apparent and accessible. For example, our assessment of the future performance of a general in combat is analogous to Newton's intuition about a problem—well before any mathematical demonstration, and certainly long before there was adequate observational knowledge of the nature of gravitational interaction.

Judgments resulting from an abundance of information are conditioned by thousands of alternatives. Decisions then need to be made much more frequently. Car designs, for example, might be modified every month rather than on the year-by-year harvest cycle introduced in the still-agricultural first half of this century.

In the future, there will be continuous anticipated feedback to the producer from the eventual customer. The "end" of a transaction will be defined far closer to the real end. A new degree of control, by the hours rather than by the months, will change our notion of a finished product. At present, making changes in a major industrial process is like turning a super-tanker. Like Alexander Pope's spider "living along the line," all parts of the process will be inter-feeling. We will not be con-fronted by impossible and unanswerable demands or by unap-pealing but irreplaceable finished products. Competition will then truly become "a form of disguised cooperation" as com-petitors probe one another for the best, most adaptable niches.

At least three other remarkably new options are offered by information abundance. First, not all of this wealth of infor-mation will be accessible. Privileged access will still be neces-sary—for military workshops, for company laboratories, and especially for the database of our own PCs. Second, we will not be able to assume that we know how something really hap-pened, or how a decision was made, even if we have access to all the information. A substantial amount of information will certainly be designed to offer "rational" explanations after the fact. Third, there is the issue of timeliness: is the information the latest and the best, or is it left over from a bygone era?

The many time differentials and bottlenecks induced by this very richness (or abundance) of data can be offset by the com-puter's sleeplessness and its prodigious outreach capacity. It has already begun to redefine the work function as well as the workplace. The increasingly desperate need to establish the critical path of relevance through the growth of raw informa-tion makes the processes of sharing information as important as its primary identification. The computer is much more an architectural device than a desktop or a company tool, because it can build the office into the global city. For everything but teriyaki, the Tokyo division of my company is effectively down the hall from my desk in Washington.

Bill Johnson (in chapter 5) and Al McBride and Scott Brown (in chapter 6) examine what this pace entails for organizations. For example, we still carry bills and checks in our wallets as we get closer and closer to an electronic-funds society. Fortunately, banking is run by people who move deliberately. Yet the com-puter's inroads are likely to make banks increasingly dependent on expert systems to implement policy. The judgments of a lending board, greatly enhanced by industry-specific "outside expertise," will sit on an officer's desk. These packaged judg-ments could become a bank's most proprietary asset—although perhaps not a greater guarantor of success than the present decisions made about Third World and real estate loans.

According to one old definition, the most successful machine is the one with the fewest moving parts. So it is with information. Information processes are most successful when they need the fewest possible instructions, or when they do not need to go outside the network. The more sophisticated and comprehensive the level of assumption that can be built in, the wider the range of combination, the subtler the possibilities, and the greater the chance of transcending the sterile given in favorable of the breedable.

Only 80 years ago, before the creation of the Federal Reserve, New York credit was likely to be at a premium of several percentage points in, say, St. Louis. Networking is now transforming our ideas about wealth and credit even more powerfully. On-line transaction processing has revolutionized the movement and accountability of funds, as it has revolutionized other business sectors. Instantaneousness is still changing the notion of what it means to have a greatly enhanced power of decision to buy and sell. Instantaneousness is also changing the location of the sources of such power.

Wealth can now be shot around the world at the touch of a key. We can buy pesos in Tokyo faster than our voices can carry across the floor of a trading room in Chicago. Our new speed of access to our wealth, and the range of disbursement (a planetary real-time market), are themselves multipliers. They alter our sense of well-being. The ranges of possibility—in this case, access to a deepening of the market—are indices of beneficent power.

"Networking" may even be too dated a term for the flow of these great rivers of information into the sea of data. The transmissions will be so continual that they will be much closer to self-developing biological processes than to preset mechanical patterns. The noun "network" makes passive the possibilities that naturally arise out of continually renewed, mutually energizing relationships as information begets information. In the future, we may talk of "a technology environment" rather than of "a computer system," much as we would discuss a complete ecology rather than a given species.

In raising our arms, we are unaware of the neutrons firing or the muscles contracting. Johnson addresses a similarly enormous and complex process which can be called upon without conscious engagement. We are moving toward a situation of functional omnipresence, another step along the road to the information organism. "My kind and foolish comrade/Who breathes all night for me," says A. E. Housman of his heart and lungs. The network we seek is no more to be switched off than is our nervous system or our water supply. It will be cheaper and more efficient to keep on interacting, to keep growing without suspension, than to switch a machine on and off for specific purposes.

Computing and communications are re-creating worldwide marketplaces—*agorae* open to all kinds of meetings and audiences. Such marketplaces maximize and encourage willingness, not just equality. The variable will be choice, weighted by the preferences and commitments of those who enter any given place of electronic assembly.

Some people will always be in the electronic marketplace, just as today some people are always on the phone with their feet on the desk. Others will draw on their inner resources (like Leibniz sitting in his chair for days at a time to hammer out a problem), using their access only to economize a repetitive process or to ask for specific information.

Troubling externalities will always surround the organizational contributions of computer technology. Visions of a "self-managing environment" achieved through computing and networking sound too pastoral. In the wide-open world provided by computers, there will be at least as many possibilities which, unhindered, will bring danger and destruction. J. Robert Oppenheimer felt drawn toward creating the atomic bomb because it was "so technically sweet." What other "technically sweet" mischief lies within the possibilities that our authors discuss?

### The Image

The third part is titled "Knowing What Is Known" because these chapters confront the implications of barely knowing the structures of what we perceive. Until our lifetime, the visual image was something set. Roger Levien, Olaf Olafsson, and Hal Langworthy show how we will be engaged by the truly protean imageries moving toward our sensoria. Before the arrival of cinematography, around 1890, people believed that an image was a "thing" rather than an element of a sequence. Today images are synthesized and modified; they are fluid activities.

How then does one read an image? Is "image" the word for something closer to a waterfall or a cloud formation than to a painting?

The perceiver is still more complex than that which he perceives. But what will happen if the data we transmit become as complex as ourselves? This would be a radical convergence. If an artificial environment satisfies the sensory tests brought to bear on it as well as does the original it replicates, is it reality? In classical legend, birds tried to eat the succulent cherries painted by Apelles. What might be the nature of identity if we were to approach the ability to create such indistinguishable artificial environments?

We want to have the minimal number of material objects in transit, and we want to analyze things that are not directly accessible to our sensory comprehension. We also want more than passive relations with our data systems. Shouldn't the library call and tell me that otherwise uninteresting data on crop levels outside Novosibirsk have just arrived—data that, by affecting the price of the ruble, alter the prospects of my technological joint venture with the Soviets in Santa Clara?

The range and the nature of the user's needs should always be anticipated as the computer correlates them in ways the user has not considered. The computer will soon separate coherent information arrays from the background disorder, and collocate entities of analogous structure even if the user had not discovered the analogy.

A library, however, has no real-time access. "Library" may be another historically conditioned term that is dissolving as we seek to give it adequate meaning. We can now pull books out of the air. Yet the definition of "book" will become puzzling once authors continually revise their work, like a newspaper, in light of the latest research. At home, we may want to update our encyclopedias almost daily. Just as the learned journal came into being 300 years ago to accommodate faster-moving data and even more specialized needs, so we now face dynamic changes in information dissemination.

The dissolution of obstacles to information makes me wonder about the implications of going on a picnic in Vermont, hooking up my modem, and having the Library of Congress leap by satellite from Capitol Hill to my side. It also opens long vistas of the crystallization of world knowledge—for example, if all data on diseases and trauma were stored in a central "library," one could detect correlations that had never been noticed. Another long-range possibility is the marshaling of all geodetic data into one facility with the power to amplify a myriad of clues, from the flight of birds to echoes in pipes, to warn of earthquakes. Anything the learned professions need to consult will soon have to be in electronic form: "You mean, doctor, that you didn't check?"

Information abundance is disastrous without information sophistication. Civil servants, for example, tame their political "masters" through such abundance, drowning them in memos. Too often, we think of information technology as a means of doing all-too-redundant calculations rather than as a means of maximizing the pooling of information and examining what is in the pool. We do well to remember that civilized humanity had looked at the weather for 5000 years before Luke Howard categorized the clouds, barely a century before his great-grandson Roger Fry formulated the doctrine of significant form. Billions of clouds had floated by, uninterrogated.

Superior imaging stimulates the imagination. How will minds react to the sight of the future equivalent of Apelles' cherries? Will it be possible eventually to transmit pathology specimens with molecular accuracy, or to enter designs directly into the patent archives?

Once Henry VIII dissolved the monasteries, early in the 16th century, most of their archives were sold to grocers and soapsellers to wrap their wares. Anything on paper is similarly vulnerable. It is also prone to misinterpretation over time: the Oxford English Dictionary has a long list of words created by mistake, not meaning what the learned thought they did.

The baked-clay archives of ancient Sumer, as Roger Levien notes, are almost insupcrably enduring. They are certainly the oldest documents uncovered. But a Sumerian document was entirely a single undertaking, not reproducible. It could not circulate, it was burdensome to store, it could break your foot if dropped, and it was difficult to write. Nevertheless, it was a gigantic improvement over memory for organizable, nonritual material.

Writing on paper followed. It offered amenability and transmittability, but it was far more perishable. Most of what we know of Greek literature was found in 19th-century desert excavations, the bulk of the treasury having rotted in wetter climates.

Printing was first valued primarily as "the art of preserving arts"—conserving and multiplying documents, rather than disseminating them. Renaissance epigrams often spoke about these overriding powers of conservation. Public announcements, which are of fleeting interest except to the historian, continued to be written on walls because of the high unit price of paper.

The history of printing over the last 500 years has been a tale of the diminution of the effort required to establish, disseminate, transmit, and record information. Now we don't always need to write a book to see it printed; we can just write the instructions. Human consciousness is increasingly operating at many removes—technical, geographical, intellectual—from the consequences of its initiatives. More and more, data are machine-sensed, machine-recorded, machine-accessed, machine-replicated, and machine-resolved—without human intervention at any point.

How far should we distance ourselves from all the necessary processes of acquiring and acting upon information? Despite the speed of the computer's response and communication, there will still be some decisions that will demand human judgment. How much responsibility should be delegated to our "faithful friend"? We may want the computer to accelerate decisions made in chess. To what extent do we also want it to accelerate decisions made during war?

Wall Street's controversies over program trading come to mind, as does the bold decision of a 26-year-old guidance officer to override a computer's command to abort the eventually successful descent of the first manned lunar lander just before it entered the "dead man's zone." The more complex and wide-ranging the information systems we create from now on, the more difficult it will be to anticipate the costs of what will happen when man himself intervenes in the process.

The symbolic conveyance of formalized information itself has fascinating antecedents. Cicero used some form of shorthand, yet the symbol for long division (one example of formalized notation) did not arrive until the 17th century. The first typewriter with raised letters appeared a century later, and then carbon paper (developed for the telegraph form) came to be used for duplication.

Before the Fourdrinier process introduced cheap paper, the only pictures in the houses of the poor were the heads on coins.

Socrates probably never saw a picture other than one on a vase or in a public building—certainly not in his house. Pictures were later seen in churches, and painted at country fairs. The appearance of framed prints, in the 18th century, must have enhanced the visual richness of life tremendously.

The copying of plans, maps, and charts began relatively late. The problem of copying pictures was a drag on reproduction until the invention of xerography. Xerographic copying redefined the category of people seen as having something useful to say, just as printing redefined the category of people capable of being taught. It also has been a check on large-scale dishonesty: there will always be something in somebody's file, as the lawyers warn us with gleaming eyes.

Copying did more than just democratize access. It also encouraged feedback—people began to write on a copied document and return it rather than type a reply on a blank sheet. The fax machine is another great inclusionary device—people on different continents can now send written communications in real time.

Technologies can substitute for institutions, from the scriptorium to the newspaper. Access to information increased with the appearance of libraries, with the numbering of pages, and (especially) with the systematic versification of the Bible in the 16th and 17th centuries. The slide rule (which is as arcane as the abacus to our children) created its own history of active, written, more intensely organized and analyzed material.

There has been a steady increase in the compactness and the accessibility of documents, from the Sumerian brick carrying fifty words to the optical disk, one hundredth as heavy and carrying many times more information. But computers and the human mind both usually possess more information than they can access efficiently. In each case, the capacity to store information vastly exceeds the ability to apply it efficiently. My company, for example, has been responsible for planning how to enter all the documents of the Washington transit authority on optical disks. The mandate was to give a track worker with a sixth-grade education access to any construction blueprint while deep within the tunnels of the Red Line. The idioms of the information era—including "accessibility" and "empowerment"—came to life when I realized that neither an official in the document warehouse nor even the track worker's foreman had to be in the loop. How much longer will it be before the display in the tunnel *anticipates* the needs of this track worker?

There is a convergent aspect of facility. Instruments as well as surfaces improve. Mechanical difficulty diminished steadily from the stylus through the quill, the iron nib, and the fountain pen. No longer does a "typewriter," as in the 19th century, have to wear protective clothing because of spraying ink; no longer does a typographer appear drunk because of lead poisoning.

Specific documents are becoming much more transparent: we now have easy access to the information that lies behind them. We call up footnotes, search Lexus, and ask for additional material. We are connected to, and able to interact with, a huge population involved in computing or word processing. Systems should be organized to interact with and to scan ahead of us, and to bring to our attention possibly relevant material. Such developments are implicit in the notation of music and in rhyming dictionaries. Will we be told what kind of music a poem may have been written for? Books, however, have always been a rich and grossly underused technology. Will computers be different?

### The Interface

The explosion of computer technology challenges us to imagine what will come next. We assume that almost anything is possible. We are not surprised by what is achieved.

Buckminster Fuller described his reaction to news of the Wright brothers this way: "Our parents pooh-poohed, but we children always knew it would happen." A generation ago, the things described in science fiction were dismissed as childish fantasies. Now they are inescapable. Many computer scientists and analysts, including Alan Kay and Timothy Dickinson, are keen students of science fiction, and their own writings are sown with reference to Isaac Asimov, James Blish, and Arthur C. Clarke. What will come next? The answer suggested by another Clarke's Law—anything that the most eminent living scientist of the last generation declared impossible—has often proved right.

In the past, ideas often preceded technology in a frustrating way. There had to be an inflow of many superficially unrelated streams of talent before genius could fish a relevant achievement out of the technological sea. Sir George Cayley, for example, sent his coachman down a hill in a glider in the 1830s, but knew that powered flight awaited the invention of the internal-combustion engine. The engine could be envisioned, but the engineering remained too complex. Lord Byron's daughter, Ada Lovelace, invented programming for Charles Babbage as he developed the idea of the (necessarily mechanical) computer far ahead of the gear-tolerance technology of his time. The great philosopher and linguist Charles Sanders Peirce was hamstrung by engineering difficulties as he casually described an electric calculator around 1895 in the margin of a letter. Today, however, technology often precedes imagination.

The search for new technology also brings surprises, such as when scientists checking on extraplanetary interference from background frequencies heard "echoes" of the Big Bang. Many insights are in fact transplanted answers to other questions. As we ask larger questions, the answers can have shattering effects in other categories. For example, in the 19th century the challenges posed to the biblical account of creation by the theory of evolution and the new estimates of the age of the earth were more in the moral arena than the scientific. Answers and insights can lead to new and distressing perspectives, such as when Madame d'Houdetot saw the first balloon and murmured: "Soon they will discover how to live forever—and *we* will all be dead."

In chapter 10, Tim Dickinson and I ask how technology can awaken human possibility. In the Athens of the 5th century B.C., it was worth recording of even a person of eminence that "he owned books." Until recently, having a pool of information around one meant having a library (or a collection of carved tallies or knotted thongs), or having as advisors several elders who had memorized a million lines. What will be possible when everyone is in such a pool, like fish in the sea?

Classics professors from Harvard, Bowdoin, and Boston Universities and Pomona College have already developed a software package, Perseus, that transports ancient Greece into the information age. Perseus uses graphics and video to teach Greek history, art, philosophy, literature, and religion. The texts of Greek plays are recorded on a read-only optical disk, which also includes maps. This ingenious venture may carry us a bit closer to its subject, but it does not help us understand the strange congruence of extraordinary intellectual power with the most brutal forms of chattel slavery that was "the City of the violet crown."

Computing and electronic communication become demystified as they pass into habit. Only a few years ago, working at a computer was thought of as something done by gods or nerds, but not by the vice-president for sales. Today we calculate, compose, and communicate with computers while perhaps chatting with colleagues and spilling mayonnaise on the keyboard. There is no longer a shaman class of white-gowned attendants standing between us and a glass-enclosed computer room. Computer expertise becomes naturally democratic as calculation, composition, and communication increasingly become subconscious activities, like talking while practicing the piano. But how well is the computer soaking into society? Certainly not well enough when the Secretary of Education can find no improvement in literacy, let alone in math and science, over the last decade.

How, for example, can it improve education? School is hardly the most dynamic part of a child's existence. Most education today is not imparted within school walls anyway. There is even an element of rational choice when a child skips the homework assigned by $20,000-per-annum teachers for the polished efforts of million-dollar producers and video stars. The learning environment needs to be energized. But Dickinson and I warn that the computer's role in such excitement might very likely prove counterproductive.

Nor can we anticipate the consequences as more and more adults expect their skills not to ripen with experience but to be discounted at faster and faster rates—to feel internal intellectual defeatism, to know that the next levels of skills are "too hard," to face the complex of attitudes that makes it so difficult to teach most people languages once they know what elaborate structures they are.

The computer is not a unitary phenomenon for a child or anyone else to "get accustomed to." We don't really know what a computer is until its multitude of implicit capacities have been elicited as solid applications. (We wouldn't know what a paint brush could do until we had seen both a Rembrandt and a Larry Rivers.) We have a poor sense of the connectivity of possibilities as one breakthrough multiplies another.

At what point, for example, will computer consciousness induce mathematical confidence or mathematical laziness—dynamic information outreach or reliance on unexamined programmers? Fear is a highly general overriding mechanism, and a great mobilizer. Can we program in "Ghosts are real"?

Newton conceived calculus without envisioning the oceanic range of what it made possible. Doodling with a computer is now better than doodling with a pen. One can certainly speculate with a pen and paper, but computation increases speculative power. When the zero was conceived (around the eighth century), and numbers could be arranged by magnitude, the creative consequences were astonishing. Computing ultimately offers powers more like those of a sculptor's chisel than those of a pencil, freeing the user from a single dimension. What might happen when nearly every child in the country has immediate computer access to nearly every other child in the country?

Lee Hoevel, the author of chapter 11, works on the human to-computer interface for NCR. He seeks to enable us to develop a second nature of symbiotic resort to our computers, and to enable computers to interact with the whole galaxy of information forms. We are undertaking things of such complexity and power that they raise questions about the user rather than about the system. Meanwhile, the flow of thought is accelerating, even by the standards of the end of the 20th century. Thought so greatly streamlined will entail internal as well as external consequences.

The computer performs mundane tasks that would otherwise be performed by whole cadres of neurotic pedants. But we don't know that anything is really and merely mundane until that has been rigorously proved. Which human brilliancies can we expect machines to approximate, or to synthesize some aspects of? Can the machine achieve a human level of problem-solving abilities—the economical, imaginative leap rather than the eliminative (and possibly reductive) plod?

The complexity of human communication is mediated through many different means. We don't know enough to assert absolute incapacities, no matter how many improbabilities we may divine. There are already vastly more possibilities than we recognize—and how many are yet unrecognized? In chapter 12 Bob Lucky goes well beyond year 2001 to dream of translating telephones into communications surrogates that will act as our agents, and of telepresence, which will carry all our senses to observe environments beyond our bodies' power to endure. But what do we ultimately ask of the capacity to communicate?

Here we encounter ramifying social and psychological ambiguities. For example, it may be possible for a house to provide for contact with the outside world. The physical need to go to special places to do certain things is steadily diminishing. Yet so much human interaction has always occurred through random encounters. What happens when going out is an activity of choice, and indeed what is "going out"—or "staying in"—in a world of planet-wide computer interrogation? Just as our notion of the structure and magnitude of the physical world is changing, so too is our notion of what our social circles embrace.

Until recently, people experienced information as fundamentally disconnected; if it was to be organized, it had to be rendered into verse, or alphabetized, or tabulated. Only now can it simmer in an interchanging matrix. Today, not only is information always available in oceanic quantities; so also is programmed instruction.

Physically, this is a knowledge-hostile world. Signs are obscured by barriers, and signals are drowned in noise. But there has been a perpetual pushing back of loneliness with each surge of communications technology. Farmers' wives no longer go mad on the prairie from sheer loneliness. The path from writing to broadcasting to computing and beyond dissolves the isolation of individual life and intellectual activity.

### Stormy Currents in the Information Sea

Economists are wrestling poorly with the problem of how to value the productivity these technologies have introduced, just as business is having trouble balancing the demand for integrated information systems with the need for decentralized information use. Service productivity and its derivative possibilities are chronically underestimated.

Much of service productivity is necessarily and properly external, and thousands of externalities are buried in services. For example, depositors can ascertain their bank balances over the phone without driving to the bank, or use an ATM rather than fidget in line. Increasingly sophisticated weather forecasts reduce the chance of being stranded in an ice storm on one's way to work. None of these savings of time and energy show up in productivity tables, nor is the proliferation of personal computers and video recorders reckoned into the index of living standards.

Service efficiency extends freedom of choice. Such enlarged choices should be seen as increases in amenity for the entire world, although they do not show up in statistics. A general improvement in efficiency is at least as desirable for the Nobel laureate as it is for the taxi driver. But how do we credit a given technical achievement as "productivity"? How to calculate the globally rewarding export returns of America's unique "big science" projects? Landing men on the moon and mapping the human gene, which are otherwise costed to zero, energize other countries, whose own research achievements are all too often locked up as trade secrets.

Computers are the key instruments of flexibility, from testing the market to supplying the market. This is the epoch in which just about any pressure on real resources works as an incentive to computerization—a more powerful analogue of earlier processes such as substitution and time-and-motion analysis. Yet too many economists still treat the computer as another real resource instead of as the master filter of real resources. It is a multiplier of individual effort, an interrogator and a cross-referencer of different skills rather than a replacement for them.

Anything asserted to be worthwhile must maintain its validity through continuous testing. Computers are increasingly offering working alternatives to the interesting but scarcely operational academic skills of the past twenty years. Their powers can be demonstrated—and challenged—day by day. The computer, which once threatened to delegitimize the middle class through unemployment, is now one of the main instruments of its contribution.

My students tend to take networks for granted, as a condition of efficiency—as a maximizer of possibilities that let them be participants free of institutional hierarchies. In business competition, proven operational usefulness counts more than the implications of "an education."

### Management and Technology

The power of information to make itself dynamic is the ultimate intangible asset. The denser the stock of information and the wider its dissemination, the more we approach what economists call "the perfect market." Much of the unexpected flattening of the business cycle can be credited to the superior micro-adjustment which computing now affords for the control of inventory.

A service economy cannot neglect manufacturing; if it does, it moves faster and faster toward obsolescence. But only manufacturing in which information is central will thrive in the next century. America is not so much an information-based society as it is a society transfigured by information, the idea racing ahead of each faltering realization. The freedom to be so structured—which means to be so powerfully flexible—is predicated on the free movement of goods and services which are most densely industrial—including computer manufacturing.

In business, owing to the speed of change, an increasing amount of technical (and therefore policy, purchasing, and personnel) decisions will be made by people not expert in technology. They need not be engineers; however, they will need to be educable in overall trends and directions, and they will have to possess a strong grasp of the resource fundamentals that underlie technology questions.

The word "management" is noticeably dated. Management as a science has been a conscious, separable discipline. Now we are reconstituting a marketplace in which there is much more intensely continuous transaction. Instead of price initiatives stemming merely (and likely dishonestly) downward from the center, we have endlessly resonant sequences of feedback over the whole surface of the interaction. And they are much more extensive than was expected. We are trying to raise the level of specifically conscious functions—to substitute sensible, repetitive, mechanically operable transactions for the all-too-human responses of the gods in the head office.

"Management," however, implies that certain aspects of these transactions are uniquely "human." But if, in a million-component network, "management" has to imply an element of inarguable command, we must soon confront an environment so information-rich, so swiftly queriable and responsive, that pure "command" is more like inflexible emotion and less and less like useful reason. The free-market vision of a world more richly developed by rules rather than by laws is a vision of a world in which information is harnessed more and more. It is a world of cooperations rather than commands, of worthwhile participations rather than predetermined submissions.

"Store manager" and "supermarket scanner" are valuable lin-guistic fossils. Most everything in a current "store" is visibly part of a swift flow rather than a slow stock. A "store" is a transaction point, not a warehouse. We move ever closer to a set of inter-active cooperatives serviced by a common infrastructure, which exists only to pace and lateralize the demands of the coopera-tives, not to enforce the reverse.

The differential of success among societies, and even more among the subset of businesses, lies along the curve of worth-while reciprocity. Quality depends upon the participants. The relevance and the relative importance of each participant change throughout each transaction. Societies that appear purely subordinating, like Japan, yield on closer examination great richness of collaboration. Respect for competence is at its lowest in societies, such as the Soviet Union, that have sub-verted democratic forms with bureaucratized, premature, final decisions.

For nearly 200 years, military general staffs have been the models of "task-focused groups." We have long known that people perform best when involved in pursuing related but dif-ferent things as members of a team, rather than when viewed as interchangeable elements in a large and inefficient pool. The task-focused group is not a transcendent union of administra-tion and technology, as management consultants would have us believe. But such groups can often be enhanced through tech-nology. They can also end up being task-*generating* groups, be-cause beyond a certain level of complexity they can raise questions more worthwhile than the fulfillment of their pri-mary tasks.

An army's approach to waging war is remarkably similar to the guidance which Bill Johnson offers to corporate executive officers in chapter 5. In each case, tasks steadily shift to rede-fine themselves in the execution. This happens because of the ever-increasing discovery of the complexity of the objectives and of the participating organizations. Bonaparte, who may just possibly have known, laid it down thus: "He who says he has made war with no mistakes has made but little war." The networked CEO is going to make tactical errors, just as he did before, but he is in a position to receive negative feedback and to recover at the strategic level more quickly.

The computer's pervasive effect works endlessly to force "au-thority" to prove itself. For example, the hierarchy epitomized by the unique position of the CEO is demonstrably hard to define in increasingly complex institutions. Such terminological distinctions as "end user," "developer," and "information system manager" age as we examine them. Yet there is a bleaker side to these technology-vitalized renewals of organization: universal access to participation in an operation or transaction can be a trap.

We all want to explore another recursion, just one more study, before submitting our work. There is a tremendous distracting potential inherent in the immense growth of communication. Moreover, the horizonless interaction of all these happy cooperators can become highly satisfying to the network's designers and unnecessarily expensive to its employers. My California office, for example, calls me up in Washington a half-dozen times a day. I would rather receive only one or two concise communications. So much contact has a way of becoming "wallpaper," as McLuhan called piped-in music—something heard but not listened to, an emotional rather than a cognitive fact.

Teachers (or systems) setting out to induce hierarchies of awareness will still be needed to establish the whens and hows of what work is to be done (or not done—e.g. cutting down a rain forest). In talking about the erosion of hierarchies, we cannot assume that most of the tasks presenting themselves will command an identifying loyalty among those who are expected to do them.

The perfection of information in a given form will not solve problems of time-discipline, motivation, or capacity to establish priorities, or other problems of organization and cooperation. The less accountability is enforced, the more difficult it can become to assay work. What is needed is not to just improve job efficiency, but to reconceptualize tasks—yet another management frontier for symbolizing intelligence.

Aside from the questions of oversight by computer, and whatever other intrusions break out in quarrels between organized talent groups, it may be very difficult for firms to account by specific units of time and product when it is so easy for workers to melt into the organizational flow. The more complex an institution and the more autonomous its participants, the easier it is for people to appear to perform X while actually avoiding X or doing Y. The president's chief of staff, for example, may find it surprisingly difficult to be absolutely certain of the real reasons for lieutenant colonels on the National Security Council staff to be flying around Central America or Southwest Asia.

Hierarchical structures may, to date, exercise the most healthily economical (if not the most creatively provocative) denying and disciplining function upon their organization—a function somewhat like that of the Office of Management and Budget or the Treasury Department in the US government. The more we open up the possibilities of communication, the more room there is for the bad as well as for the good to percolate. The larger and swifter-growing the forest, the more rotten leaves can be hidden. The endless subterranean interactions in Chicago's Mercantile Exchange and Board of Trade come to mind.

Computers ensure that more and more initiative, both absolutely and relatively, will move outward to the "marginal" and the "subordinate." There will be more consequences than anyone has worked out, whether for the corporate executive or for the biological research team. An independent design firm, for example, might deal directly with the factory for which it is shaping a saleable object. Companies of all sorts may increasingly resemble advertising agencies or law firms, traditionally allowed to have recombining gatherings of rising and falling talent. Authority will have a vital but receding place: it will still be necessary for someone to enforce the consequences of actions which everyone deplores but no member of an assembly wants to be the first to prevent.

The CEOs of Corning Glass and Levi Strauss, two innovative companies, are enthusiastic about such 21st-century management concepts as "the company without walls" or "the global network organization." But these visions must be reexamined. What form will the necessary authority take as people at the hub diffuse information to outlying partners, associations, and consortia (and in exchange for what)? Here we must confront not just the nature of the organization but that of the system of production. An entrepreneur's ambition of building a 21st-century Noble House along the Pacific Rim could become problematic as he tries to exercise command over a realm which is culturally and intellectually more diffuse than that of its founding capital.

Corporate as well as national cultures are being transformed faster and more extensively than they were by printing and radio. Even France's insular and highly evolved culture has been energized from outside—by business and by "Franglais," among other influences. Among both companies and states, technology is shaping each entity to a startling individuality. No management formula or precedent presses itself on the CEO or the politician. Vitality must be enhanced, irrespective of consequences for the corporate or national culture. It ensures that diversity can survive and proliferate. What would the departed Walter Chrysler have told the disoriented executives who hired mediums and held seances to seek his guidance? He would have fired them on the spot. No formula works for everyone, and this is as it should be.

Computation has already transformed organization and management. Speculation about its imminent future must take into account attitudinal friction as well as technological momentum. The thirst for elementary certainty within the operational (or "corporate") culture is all too likely to outweigh the need to create an army of limited, objective specialists. Americans particularly love to delegate responsibility to experts, many of these "experts" so certified only by their clients. Unsurprisingly, many American businesses are still increasing the organizational levels and the numbers of their "managers." As Parkinson's Law suggests, a surplus is all too likely to be spent by its creators before those for whom it is created can learn of it and put it to work.

Cross-functional cooperation, on the other hand, is part of the Japanese secret. Egos are collectively warmed in corporate-size research projects and on the shop floor, although senior executives are anything but collegial in resource management. Similarly, German institutions are superb at focusing intense cooperation under a few commanders who are very hard to challenge.

Consider all the "consulting" in and to US companies: the office psychologist, the auditor, the ergonomist, the team of MBAs from McKinsey, and even the technology forecaster. Such interaction and collegiality is likely to evoke the weaknesses as well as the strengths of the culture whose sensitivities (and mere insecurities) these consultants set out to anesthetize. This interaction is at least likely to give these strengths and weaknesses new prominence.

Technology is sporadic. As in the history of optics, it may take thousands of years to realize a theory. Technology opens a wider spectrum of possible outcomes without guaranteeing them. It radically challenges organizational assumptions left over from the heyday of "managerialism as a science." Many corporate leaders are still paying the price of having been forced to assimilate business-school management in the name of science. The notion that a meta-discipline can be imposed on human possibility will, in the decade ahead, be shown ever more vividly to be a perilous combination of arrogance and inflexibility.

### Technology and Internationalization

Twenty years ago, the cliché was that IBM was not a competitor; it was the environment. HAL, a machine so powerful that it could turn human beings into its own tools in Clarke's *2001*, derived its name from a dancing sequence of the letters I, B, and M. IBM once determined the computer's capabilities and then set the pace and the terms of competition. But a great culture is more than its institutions.

The notion of compatibility has evolved, as has America's planetary example. The heroization of the corporation is obsolescent. The developmental history of the international order shows a certain parallelism with that of computer manufacturing. Whereas the world of 1970 was still dominated by the US and IBM (spearheaded by Dick Watson's IBM World Trade), the 1990s emphasize national and corporate access, functional equality, and wide dispersion in a world boiling with initiative.

Today's map of relative power is likely to look two-dimensional in comparison with that of tomorrow. Yet it is likely that the 21st century will be an even more deeply American century, as Seizaburo Sato, an aide to Japan's former prime minister Yasuhiro Nakasone, predicts. This will happen to the extent that the United States remains at the center of the international system and continues to carry the institutional values it created. A country, like a company, needs to be strong among the strong, and strong (not escapist) among the weak and resentful.

Any future is a combination of technical possibilities and relations with strangers—in our case, high technology and internationalization verging on the cosmopolitan. The physical trends which the authors address may be relatively clear. Yet discussions of the world order of the next generation—the larger arrangements that will shape these trends—have recently been prone to bad analogy, bad analysis, and plain refusal to recognize the international framework. For example,

Kenichi Ohmae, director of the Tokyo office of McKinsey & Co., denounces the United States' "mismanagement" of the world economy and asserts Japan's "right to share world leadership," and Yale's Paul Kennedy has likened America's "imperial overstretch" to the hubris that brought down history's profligate "great powers."

Will Babylon's next successor be Armonk, Blue Bell, or Cupertino? Hardly. America at its best is history's great exemplar of unthreatening hospitality. It has neither exhausted nor maximized this unique role.

The US trade deficit has been Exhibit A for what the Germans call *Gewaltsmenschen*—political and academic "crisis men" who thirst for emergencies in which to flourish. This still not adequately quantified trade balance is due not to incapability or indiscipline but to the fact that the people of America's largest trading partners, Japan and Germany, have been individually—and collectively and geographically and culturally— eager to plow their earnings into expansion and acquisition in America. This shows an enduring confidence in their principal trading partner. If we Americans want a dynamic world economy, we must choose one form of imbalance or another. The least constrained economy does well to arouse the tighter societies of Japan and Mitteleuropa.

If the United States were to accumulate a sizable trade surplus, it would suck oxygen out of the less generally confident world system. A trade deficit is offset by inputs of capital. Capital is attracted by America's constitutional and social stability and its labor flexibility, which are all centered on the liberal international system. Inflows of capital tend to finance and be financed by a short-term trade deficit as long as the United States remains the best long-term place to invest.

The "crisis men" also detect evidence of decline in the fact that the United States is now the world' largest debtor. Irresponsible comparisons are drawn, using figures that reflect a net asset position in which the negative differential is not only rather easily financed debt but eagerly invested equity, priced above market value. The debts of the United States are mostly in its own dollars, rather than in currencies earned through international trade. As Adam Smith and David Hume taught us 200 years ago, a dynamic economy can flourish as a net debtor; an unreceptive economy may stifle on its own capital balances.

If there is to be international economy, there must be trade-offs. And the spaciousness of American life has always required a high degree of consumption. This is a functional consequence—not a coincidence, let alone a random moral choice. Moreover, the faster assets are used, the less the impulse to save as cash (as contrasted to trying to save in terms of education or speculation). Technical change is likely to accelerate spending. It can move investment into less visible capacities, such as company design teams or computer training. Individual time and effort in self-improvement are carried as zero on the national income accounts of the United States.

Too many people, seeing a world so technologically united that a Tokyo finance house or a New York databank can work instantly with counterparts in Rome or Rio, believe that the same degree of efficient unity prevails naturally among nations. A spurious assumption of automation surrounds structures consecrated by time. That America's partners—such as Japan, South Korea, Taiwan, and the countries of Europe—should attain thermonuclear, central-banking, and policy equivalency offers more likelihood of deadly disorder and disharmony. America's moderating preeminence remains the lubricant that makes possible such a large free-trade area and reconciles such explosive forces with democratic values.

A tremendous factor working for prosperity is the reasonable certainty of general order. America cannot reap such great profits from self-sufficiency as it does from being the main beneficiary of the rich international division of labor. Lying outside Europe and Northeast Asia, history's two cauldrons of perpetual rivalry (and, perhaps not coincidentally, of economic dynamism), the United States will remain the crucial integrator of what we can call the Northern Oceans System—up to this moment, the only reciprocally creative strategic counterpart to the information sea.

The Northern Oceans System reflects the intense relationship between the United States—this greatest "island power"—and the other countries of the two basins. Fiber optics and satellites have integrated this system to a much more intensive and productive degree than was possible at the height of US supremacy. America remains the resource of last resort in defense as well as in economics, and therefore underwrites stability—predictability in a fluid world.

The American Century triumphed in the ongoing democratic revolution that overturned the ancient distinctions of class, sex, and race; in the globalization of scientific benefits; and in the awakening of an extraordinary diversity of talents among ordinary people (after the 1940s' war-weary sense of individual helplessness).

The international economy is integrated more in terms of services than in terms of goods. The more impalpables there are in an economy, the more it will be internationalized by default. You could tell where a coal seam was; you couldn't tell who was going to challenge the mail coach with the electric telegraph. The most competitive nations by year 2001 will be those that can best mobilize talent—and then adapt to it. Technological rivalry becomes rivalry for ability, or rather abilities. In the Malthusian debate, each new person can be construed as either a mouth to be fed or a brain and a pair of hands to supply others. In the 1990s, the more efficient a state's communications, the more it can envision its citizenry as likely talents.

Enhanced mobilization is one of the central consequences of information technology. The banks' readiness to write off foreign debts reflects their despair over the Less Developed Countries' organizational capacities, which they idealized at the first fat interest payment sixty quarters ago. Yet individual LDC talent is ever more welcome as it is increasingly mobilized behind its computer screens at home.

In addition, one of the rewards of working in just about any large or small US computer company, including my own, is that one's engineering colleagues come from all points of the globe and are in touch with further capacities worldwide. If America relaxes its remaining barriers against the immigration of skilled creators of knowledge, the rate of its advances in technology and industrial productivity will increase—and then the foreign-born creators will return to fertilize their lands of origin.

Japan, on the other hand, is by no means a uniform success. Its rapidly aging society has neither mobilized its female population nor welcomed immigration. Japan may have gone from being the weakest great power in 1941 to the greatest weak power 50 years later, but it has not transcended its problems of distribution and infrastructure. Moreover, there is no indication that the Japanese have taken leadership positions at the innumerable frontiers of raw innovation, however triumphantly they lead the world in the capacity to convert innovation to high-quality, high-volume manufacturing.

Japan's economic future is constrained by geopolitical fundamentals: its natural greatness is not as survivable as those of the continental powers. The fact that most of its citizenry is crowded into bowls of hills along the coast prevents Japan from translating first-class economic strength into first-class military standing—still a requirement for being a three-dimensional superpower, and even for spurring symmetrically sophisticated innovation.

Nevertheless, Japan is too often used to personify the external pressure Americans feel from the sheer momentum of history. For example, even while we tolerate criminally wasteful secondary schooling, many of us resent Japan for having the world's largest system of decent high schools, with a powerful base in math and science. Thus, to an extent, Japan is being blamed for America's realization that it no longer has a patent on modernity.

Similarly, analyses of the high-technology future generally exaggerate the economic impact of the 1992 European Economic Community. Far more interesting is the fact that the Europeans seek to achieve a central currency. Europe is the leading foreign market for US-made semiconductors and the prime destination for US exports of high technology. Yet the thunderous economic and political reunification of Germany is putting immense pressure on the concept of a tidily balanced united Europe (which, after all, had been predicated largely on the division of the most formidable power west of the Volga). Moreover, the economic integration of Europe has been achieved with and through America's presence and good will, as is evident in the proportion of European capital and talent organized through US connections—consider, for example, IBM's two European Nobel laureates for superconductivity.

At the state level, Common Europe still has a web of rules consciously resistant to US economic interests. But the story is much different at the level of international enterprise. The road from Lyon to Düsseldorf often leads through Dayton or Maynard. The computer industry is becoming all the more richly and complexly globalized in sourcing and research at the same time that US-based companies are entering into many mutual arrangements at home. There are very few pure domestic competitors anymore.

In Eastern Europe, revolution was catalyzed by telecommunications, copiers, and PCs. These states now must place a premium on efficiency to compensate for a generation in which the denial of individual participation in the market as well as in politics led to stagnation, resulted in a collapsing housing stock, a poisoned environment, a smothered price mechanism, a decaying infrastructure, and unstable currencies. Much efficiency can be encouraged through computation in financial management or the loosening of labor rigidity through new means of access. Countries no longer have to plod through the schematized historic sequence of industrial revolution; they can go straight from diesel engines to microchips.

Technology's penetration also complements the richness in mathematical skill of such historically distinguished scientific cultures as Poland and Hungary. Initial public offerings are not about to roar out of the nonexistent garages of the proletariat; however, the impact of foreign technology in underwriting the transition of Eastern Europe will far transcend the simple but essential gift of uncontrolled, wide-ranging information offered by the faxes and phones that made it possible.

However, the United States must finance the next generation of its great technologies. Trade deficits are bad to the extent that they undercut development and reinvestment, but (unintendedly) good to the extent that they force the world's largest economy to put aside any fantasies of self-sufficiency. Without the profits to form capital or to pay for bond issues, it will be hard to underwrite huge research-and-development programs. The United States risks fooling itself by its own dynamism. No additions have been made in the last ten years to the list of great postwar industries: commercial aircraft, chemicals, communications, computers, and so on. Silicon Valley and Genentech already existed in 1980. The world is moving faster. But who in America is limbering up for a leap equivalent to that from calculating machines to computers? Even superconductivity is deeply rooted in existing technology.

All our prognoses for the year 2001 are based on the assumption that America will be a player in technology—but not always the winner. Playing the game involves buying subassemblies as well as selling subassemblies, buying from the Japanese and Europeans as well as selling to them.

The economic competitiveness of societies shifts in ever more transient and complicated contexts. The higher the technology, the more unforgiving the moment. Competitiveness can change within a few months or even a few days, although specific intercultural shifts in competitiveness take longer. We are seeing the obituaries of institutions that thought they had half a generation to go. Intercorporate competition concerns the balance struck between human resources and technical capacities, an endless dialectic of improvement in work rather than the dusty warehouse of preserved jobs.

There have been trade crises before and there will be trade crises again. What is significant now is the steady shift of emphasis toward moving ideas rather than objects. Rapid innovation plays to some unique American strengths. Along the way, the great discovery of the late 20th century is that technology transcends politics and sets its own agenda.

Computers bring far more transcendent access to our world and equality of talent than revolutionaries in the early decades of this century ever demanded. The vision of social participation originally pursued in terms of war and revolution is now actually available—not as soon-burnt-out passion, but in the real-world terms of growth, productivity, and inventiveness. Nearly every objective set by political revolution has been transcended by technological improvement. Marconi and von Neumann did more to abolish the distinction between the country and the city than all the collectivizers who ever devastated the Ukraine.

### Conclusion

Several predictions from the otherwise far-seeing pioneers of the computer industry are sobering. "I think there is a world market for about five computers," Thomas J. Watson, Sr. estimated soon after World War II. At the dawn of the PC revolution, in the late 1970s, Kenneth Olsen concluded that "there is no reason for any individual to have a computer in their home." Undoubtedly some of our own visions are flawed; but although we may not always get the date right, we are pretty certain that we're always too conservative about the scope.

Several predictions *can* be offered with some confidence—one concerning an overwhelming policy issue and a half-dozen concerning trends in technology.

The United States has all the tools to revolutionize its failed schools. We do not need to go to Japan for, say, mathematical workbenches, or tools that allow visualization of mathematical formulas and concepts from algebra through calculus. These could change the way we teach, especially since more people are driven from mathematics by difficulties of envisioning relational statements than by lack of reasoning processes.

Can the United States change its bleak educational record in math and science by literally remaking its children's visions of these disciplines? Mathematics is simply too powerful an art to exclude from a going culture: It took the fall of the ancient world to halt geometry. Those who fear mathematics might as well want to build houses while embargoing hammers and saws. Students disheartened before learning differential calculus will not go forward to quantum mechanics.

Paradoxically, the new learning technologies bring us back to the springtimes of the disciplines—when geometry, say, was really "measuring the earth." When it comes to interfacing these cities of data, elementary approaches will provide opportunities a decade from now. Handwriting and notebook computing, for example, will be coupled with the input of continuous speech processing, itself achieved either through massive data correlation backed by a statistical model of grammar or through some yet-undreamed-of elegant solution hit upon by someone who is unaware of what is impossible. (Goethe tapped out the meters of his poems on his mistresses' backs, but few of lesser genius wake up, reach for their sweethearts, and keypunch "good morning" on their hands.) Once a screen or voice box is made available, computers will infiltrate American life even further, soon coming into the hands of 50–60 percent of the population. Changes in quantity will become changes in quality.

And this access can be accelerated by flat-console technology. The machine will be on the wall. Rather than reach and then wait while turning it on, the users will monitor it from wherever they are, and will talk to it amid combinations of image-quality graphics and digital data.

The neural networks now forcing comparisons to biology will break open entirely new categories of problems, embracing the whole world of learnable, repeatable patterns—the processing of speech, images, vision, and what else? These are likely to be the first approximations. Soon enough, a system should recognize the signature on an incoming fax, approve it, and file the document. Neural networking is still very much a shadowy

promise in research and development. But the capacity to enlarge what still can only be called "judgment" radiates into innumerable activities too complex for consciousness. The most
instructive practical example to date of neural networking,
however, cannot be quantified: It is the way a small child learns
by discovering certain patterns and repetitions (perhaps learning something never before learned, or not retained by the
culture).

We are driving toward a set of imaging technologies that by
the year 2001 should display objects configured in anywhere
from six to hundreds of surfaces (as prices dictate). And we are
only at the cave painting level of visualization, even though we
can now "walk through" a building yet unraised and "see"
where the shadows will fall before the ground is broken.

Another startling mediator of change at our very elbow is the
humbly worthwhile gallium arsenide. It is likely that by 2001
every house in the metroplex will be penetrated by optical fibers—not just for media but also for networking. The public
will expect more than electronic mail and high definition TV.
Increased bandwidths and 100,000-megabyte local-area networks will carry their users from the standing of farmers waiting for the mail to that of scientists at a great laboratory.

A longer-term prediction is that end-user "faces" will be moving within ten years toward a functional sensitivity to emotion
and urgency, just as end-user interfaces today can register the
pressure with which we write. We don't want to say "Give me
the directory, computer." We want a computer at which we can
scream, two octaves above normal speech, "Damn it, find me
that scrawl Jane sent last week!" The desired computer knows
exactly who Jane is, figures out where the letters are, and reviews everything with Jane's scribbles on it received in the last
three weeks (being perceptive and experienced enough to
know that I've probably forgotten just when the letter arrived).
Then the computer asks, not even with the sardonic weariness
of the well-trained butler, "Can it be you want . . . ?" It offers a
file of Jane's recent letters and their topics, and finally it displays the correct one on my wall panel for me to quote defensively back to her in our current videophone conversation.

The computer's next great use will be to reconstruct and re-
create complex systems—to reassemble environments of the
real world as paleontologists recreate lost genera, or to recon-
stitute the genetic code as linguists reawaken extinct languages.

Impossibilities will dissolve. Which of the limiting assumptions in information technology now lie dead or dying, much like those that once imprisoned computation in mainframes?

We must combine the sense of the possible with the sense of the computer's omnipresence as the most assimilated of tools—the phone and the newspaper of the electronic age, a fact of life for industrial humanity just as the car was to young Americans born after 1918. We can anticipate a planet full of people to enter, to enlarge, and to declare possibilities that are incredible only to the fainthearted.

# *The Rising Sea*

# 1

## Brief Time, Long March: The Forward Drive of Computer Technology

*Denos C. Gazis*

Sometime around the end of this century, computer technology will celebrate its golden anniversary. Chances are that its progress will still be described by superlatives and plotted on semi-log paper. What progress can we expect in the years ahead?

Let us imagine that we are very small, and let us get into a very small airplane and fly very close to the surface of a silicon chip. We will see the magnificent landscape shown in figure 1, where the bumps show individual atoms (with their clouds of electrons around them), nicely arranged in their preferred crystal formation.

As recently as the early 1980s, figure 1 could not have been produced. Some scientists became emotional when they first saw such pictures. They had never dreamed that they would live to see atoms. The pictures became possible through the invention of the scanning tunneling microscope (STM) by Gerd Binnig and Heinrich Rohrer of the IBM Research Laboratory in Zurich. The STM consists of a probe (figure 2) so that there is almost a single atom at its tip, which is brought very close to the surface of a solid. If the tip is close enough, a voltage is applied between the probe and the surface. The magnitude of this "tunneling" current depends very strongly on the distance of the tip from the surface. If we now move the tip across the surface while maintaining a constant current with a feedback mechanism that makes minuscule adjustments to the position of the tip, we can map the topography of the surface by recording the movement of the tip. It sounds almost simple—but we are talking about tip-to-surface distances of the order of one interatomic distance, and movements of tenths of that distance. Controlling such delicate movements was a major engineering feat, which together with the conceptual breakthrough earned Binnig and Rohrer the 1986 Nobel Prize in physics. The STM

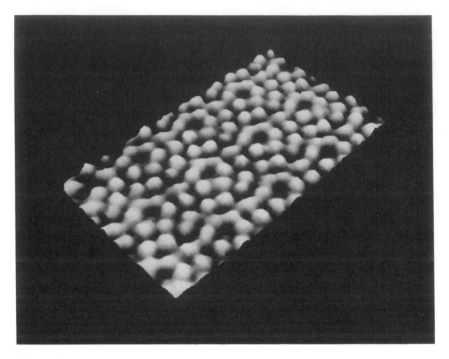

**Figure 1**
Atoms of silicon, magnified millions of times, are shown in a computer-generated image produced from data obtained with a scanning tunneling microscope (STM).

has given us unprecedented resolution in viewing surfaces—resolution which is limited not by the wavelength of a probing beam but rather by the ability to control motion on a subatomic scale.

I open with the story of the STM because it demonstrates how far we have progressed in computer technology, and also because it has given us yet another powerful tool for the investigation of surfaces, which is becoming more and more important in our quest for ever-smaller microelectronic devices.

## The Quest for Smallness

For the last two decades, progress in microelectronics has been achieved by making devices smaller, cheaper, and faster, and often by generating new functional capabilities. It is perhaps a good oversimplification to say that making things small tends to achieve all the other objectives. Smallness allows us to put a

**Figure 2**
The tungsten probe of an STM.

lot of devices on a piece of silicon real estate. Besides decreasing the cost of devices, this allows us to generate a new functional capability that exploits the concentration of computing power in a small volume. We also gain speed, because devices are generally faster when they are smaller and because we can now perform a lot of tasks on a single dense chip instead of traveling from chip to chip and wasting time in the process. I will discuss this imperative of concentrating computing power in a small volume. For now, let us agree that smallness is the name of the game in the computer business.

Ever since we started building transistors, resistors, and condensers, we have been scaling them down to smaller and smaller dimensions. We have moved from simple integrated circuits to large-scale integration (LSI), then to very-large-scale integration (VLSI), and we are now poised to move to ultra-large-scale integration (ULSI) and beyond. The minimum dimensions of today's devices are about ¾ of a micrometer. Very soon, we will move to dimensions of ½ micrometer, to be followed by dimensions of ¼ and ⅒ of a micrometer. Experimental devices have already been fabricated at those extremely small dimensions, which are invisible to an optical microscope.

But scaling is a tricky business. That is why, if you think about it, an ant does not look like a miniature elephant. You cannot simply shrink a device by a constant factor and expect it to work. Different physical properties scale in proportion to different powers of the scaling factor. In microelectronics, phenomena crop up which may always have been there but which suddenly become important in smaller devices. The net result is that when we scale down the dimensions of devices we often have to redesign them, change the materials and processes used in building them, or pay special attention to such things as the cleanliness of the air in the manufacturing facilities. After all, a thorn would hardly cause a reaction to an elephant, but it would be very upsetting to an ant.

### The Different Flavors of Microcircuitry

All computers are not created equal. They range from microcomputers to mainframes, and they differ in many ways. Sometimes I am asked: "What is the basic difference between a microcomputer and a mainframe anyway, since a microcomputer of today has the power of a mainframe of yesterday?" (In fact, I estimate that my own personal computer has about 5 times as much power and storage capability as IBM Research provided to our entire Yorktown laboratory when I joined the company in 1961.)

It is true that we have made so much progress in increasing the capabilities of microcomputers that the lines between large and small computers seem to blur. But we still have a spectrum of computers which address different needs and which are thus likely to remain functionally, organizationally, and technically different in the near future.

At one end of this spectrum we have the large mainframes, whose design is driven by the need to concentrate vast computing capability—a lot of MIPS (millions of instructions per second)—in one place. We are willing to pay a premium in chip technology for these computers in order to get maximum processing capability. The silicon chips used in the logic circuitry and in some of the limited high-performance memory (the *cache*) of these computers are a special generic family of transistors known as *bipolar* transistors. They are the most complex transistor structures, and they consume a rather large amount of electric power. This makes them rather hot, which poses special challenges in *packaging* them (that is, assembling them into a system). But they are still the fastest transistors we can get, and they will be the basic ingredients of mainframes for many years to come.

At the other end of the spectrum we have the microcomputers. The push is on to make them more compact, more economical, and substantially less voracious in power consumption than the mainframes. The logic circuitry for these computers is made up of *field-effect transistors* (FETs) of various types. The least power-hungry variety of them are known as CMOS (complementary metal oxide semiconductors). FET circuitry is also used to make all the memories for computers, large and small.

### Progress in Bipolar Transistors

Bipolar transistors do not scale down very graciously. Some of their physical characteristics, such as electrical resistance, become damagingly high. Other characteristics, including voltage requirement, do not scale much at all. The net result is that shrinking the dimensions of bipolar transistors requires extensive redesigning. Another requirement of smallness is the improved control of manufacturing tolerances. For several years now, microcircuitry has been fabricated by photolithographic deposition of successive layers of materials. This succession requires careful registration of the pattern masks, which is more and more demanding as dimensions get smaller. For this reason, many of the modern chip designs involve *self-aligned* transistors; the design itself helps draw deposition in the right places with high accuracy, eliminating some crucial registration steps. There are several iterations of bipolar transistor circuitry now under development. The intent of this development is to increase the raw speed of such circuitry by roughly an order of magnitude in the next ten years or so.

Today's bipolar transistors switch in about ½ to ⅓ of a nanosecond. There are as many nanoseconds in a second as there are seconds in 32 years. There are already experimental designs of bipolar transistors with switching speeds of about 50 *picoseconds*. Now, a picosecond is *really* small—there are as many picoseconds in a second as there are seconds in 32 *thousand* years. By the time we get to use 50-picosecond transistors, we will be running into another technical difficulty: the fact that the speed of light is excruciatingly small! We generally do not

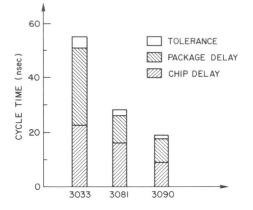

**Figure 3**
Where time is spent during computation on some large IBM systems.

think of light that way, but light travels only 1.5 centimeters in 50 picoseconds. It so happens that the speed of light is the upper limit on the speed of everything in the universe that carries energy. (We have this on Einstein's authority.) So we have to contend with a serious constraint: If we are to capitalize on those fast devices, we have to bring them very close together. Otherwise, we will be wasting on travel most of the time we had saved by making the devices fast. Every one of the large mainframes we have built in recent years has had to contend with this need to maintain a balance between the time the computing signals spend on a chip and the time they spend in the *package* (the assemblage of substrates and wiring between chips). Figure 3 shows the apportionment of the total computing time into time spent on the chips, time spent on the package, and design tolerance for the most recent three families of IBM mainframes. As can be seen from this figure, only about 30–50 percent of the total computing time is spent on the chips. It never makes sense to build a faster chip without creating a correspondingly faster package to maintain this balance. Otherwise, the overall improvement is marginal and not worth the effort.

At the system level, the speed of a machine is characterized by its *machine cycle,* the minimum amount of time during which a machine can do anything at all. Today's mainframes have a cycle of about 10–15 nanoseconds. In the next ten years, they will achieve cycles of 2–4 nanoseconds. Some supercomputers, for which performance is even more important than money, will achieve such cycles even sooner.

## Progress in FETs

As I noted above, field-effect transistors are used to make logic circuitry for microcomputers and for more and more of the midrange computers, and also to make the memories which are used for computers of all sizes. In making logic chips for microcomputers, we have long had the capability of placing all the central processing power of the computer on a single chip. The relationship of the power of the central processing unit (CPU) to its density is fairly simple: the more transistors we can place on a single chip, the more instructions per second we can have. Today's most popular microprocessor chips have about 300,000 transistors on a chip, and a performance of a few MIPS. Around 1989, Intel announced the arrival of a million-transistor chip with a performance of about 20 MIPS. Characteristically, that chip arrived just a bit early. Most experts were predicting the attainment of such chip complexity by 1990. Achieving such chip complexity involves not only the challenges of reducing the dimensions of the device but also the even greater challenges of designing the logic circuitry. Microprocessor design has been for years one of the most labor-intensive parts of the computer business. The standard rule of thumb was that it took one person-year to design a thousand transistors on a chip. It got to the point where we thought of "a thousand $X$ per person-year" as a universal constant in computing, with the $X$ standing for lines of code, transistors on a chip, and so on. It became clear that the business would grind to a halt unless we cracked that bottleneck. We could not afford chips (or programs) that were designed with such expenditure of human effort. Today, the design of microprocessor chips is done with the help of an array of design automation tools. This has reduced the requisite human labor by almost an order of magnitude. And we are working hard to reduce it even further.

Memory chips are easier to design than logic chips because their memory cells are arranged regularly in rectangular domains. This does not mean that there are no challenges in memory design. In addition to the scaling challenges, we face the challenge of improving the access time of dense memory chips (for example, by redesigning some peripheral circuitry which accomplishes this access). The industry is meeting all these challenges admirably, decreasing the minimum dimen-

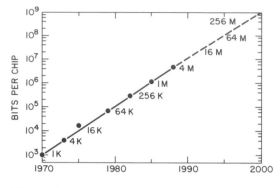

**Figure 4**
Progress in metal oxide semiconductor (MOS) memory chips over the recent past and the near future. The number of bits on a chip is quadrupled every three years.

sions of devices by about a factor of 1.5 every three years. In addition, we are able to make larger memory chips, because defects in the silicon wafers are getting sparser. The net result is that we are able to quadruple the density of memory chips roughly every three years. And progress is accelerating. Figure 4 shows the progress in memory chips, measured in bits per chip versus time, since 1970. Today, the densest memory chips used in most products have a million bits (1 Mbit). Some products are beginning to use 4-Mbit chips, which are coming off production lines at a rapidly increasing rate. The next generation of chips, with 16 Mbits, have already been designed, and preliminary work has started on 64-Mbit chips. We do not see any major technical obstacles to such progress. We can therefore expect billion-bit chips around the year 2000, with a commensurate impact on our ability to design novel applications that will take advantage of this silicon cornucopia.

## A Panoply of Beams

We have enlisted a panoply of beam-controlled technologies for the fabrication of microelectronic components. We use photon beams as well as beams that use energetic electrons, ions, and molecules to make circuits, test them, and package them. In many ways, beams of various kinds are the modern versions of the soldering gun of the primeval days of electronics. But they are more than that. Today, molecular beams are used to build epitaxial layers of various materials, with unparalleled control of the quality of the material and the thickness of the layer. Ion beams are used to do "microsurgery" on photolithographic masks, or on devices. Laser beams are used to subtract or add material in conductive lines interconnecting devices. Electron beams are used to probe the circuitry of a chip and to characterize its performance. Just about anything we can form into a beam that can interact with matter at sub-micrometer scale seems to find its way into the repertoire of tools for the fabrication of microelectronic components.

One of the key applications of beams to microelectronics is the patterning of the various deposition layers that form devices on a chip. This is done by photolithography. The minimum dimensions of devices on today's chips are just under 1 micrometer (μm). But we are beginning to approach the limits of resolution of optical projection, which is used to form these patterns in an essentially photographic process. This is because the wavelength of visible light is just a little smaller than 1 μm. And just as we cannot see small things with an optical microscope, we cannot paint small things with light. Actually, the fact that we can optically pattern features as small as 1 μm is a huge success story. A few years ago, the conventional wisdom was that optical lithography would run out of steam at about 1.5 μm. But the optical lithographers met the challenge and conquered it beyond our original expectations. Not only are they able to handle dimensions of 1 μm; they are now confident that they will be able to handle dimensions down to about 0.5 and maybe 0.3 μm. This progress has been achieved through improvements in lens design, through the use of light in the deep ultraviolet range, and through the development of advanced photosensitive materials (*resists*) that work well with such light.

Sooner or later, though, we *will* exhaust the capabilities of optical lithography. Then we will need other sources of energy to pattern resists and devices. Two prime candidates exist for lithographic processing of sub-half-micrometer dimensions: *x-ray lithography* and *electron-beam (E-beam) lithography.*

X rays can be used like light to flood a relatively large area with radiation and to expose patterns on suitable resist materials. Of course, to use x rays one has to develop every element of the process. For example, optical masks are transparent to x rays, so a brand-new mask technology has to be developed. Instead of ordinary x rays, such as the ones used in a doctor's

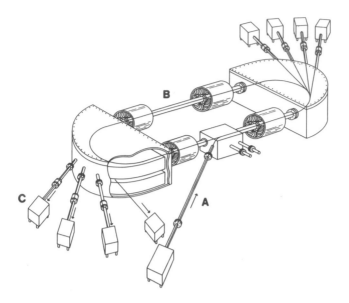

**Figure 5**
An electron storage ring used to produce x rays used for making computer chips with very small dimensions. Electrons produced by an electron gun (A) are injected into the ring (B), where they travel in a continuously accelerated path. Magnets bend the path of the electrons, and x rays are emitted around the bends, traveling down beam lines (C) to x-ray exposure stations, where they are used to write patterns on silicon wafers.

office, some special "soft" x rays, produced by *electron storage rings* (figure 5), are preferred because of their high intensity, high depth of focus, and low damage to wafers. A joint study by IBM Research and the Brookhaven National Laboratory successfully tested the concept of using a collimated beam of soft x rays to pattern silicon wafers— in 1987, fully scaled 0.5-μm devices (optimized for 0.5-μm lateral dimensions) were processed. Now an international development effort is under-way to produce compact electron storage rings that will use a strong magnetic field generated by superconducting magnets to bend the path of electrons sharply and thus reduce the foot-print of the ring. X-ray lithography is thus poised as a prime candidate for a means of producing memory chips and other commodity items in large volume in the 1990s.

"Poised as a prime candidate" are carefully chosen words. The fate of any new technology in the computer business is always somewhat uncertain. The range of applications of x rays that was considered secure in the late 1980s has already been narrowed through progress in optical lithography. Further progress in this area could defer the need for x-ray lithography, with all its startup costs and technical challenges. People who invest in x-ray lithography today probably make a mental allocation of part of the cost to insurance against competition. There is a substantial loyal opposition among optical lithographers who feel that x-ray lithography may eventually be squeezed out by improvements in optical lithography and by E-beam lithography.

E-beam lithography provides the ultimate in resolution, insofar as we are limited by the wavelength of the radiation source. Energetic electrons have wavelengths of the order of nanometers, so they have a head start in the race for resolution. They are also produced conveniently by heating a filament, and they can be easily focused by electric and magnetic fields. Interestingly enough, E-beam lithography has been under development even longer than x-ray lithography. It is already a well-developed technology, in use for some steps of microfabrication. It is not used for mass production of such things as memories, because the cost of production would be too high. The throughput of E-beam tools is more than an order of magnitude lower than that of optical tools or the potential throughput of x-ray tools. This is because E-beam lithography as practiced today is of the scanning type. A fine beam that can be turned on and off rapidly is steered under computer control to draw a pattern serially on a suitable resist material. The process can be accelerated by using a variable-shape beam that exposes as many as 100 pixels (picture elements) in parallel, but it is still too slow. Electron beams are ideally suited for making masks to be used for exposure with other radiation sources. They are also used when we need to fabricate only a few copies of a particular part—for example, a few application-specific integrated circuits (ASICs). Unquestionably, E-beam lithography will find increasing use in the future. When we talk about *nanolithography*—the fabrication of devices with dimensions of a few tens of nanometers (the decananometer range)—we can think of no other means of achieving it other than by E-beam lithography.

Working prototypes of devices in the decananometer range have already been produced. The one shown in figure 6 has minimum dimensions of 70 nanometers, and it has demon-

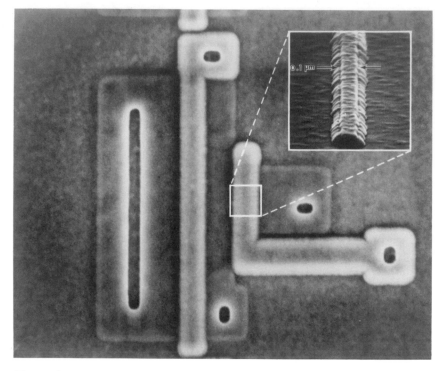

**Figure 6**
A scanning electron micrograph of transistors with dimensions just smaller
than 0.1 μm.

strated the fastest switching ever achieved in FET logic circuitry—around 13 picoseconds. These prototypes were fabricated
at IBM's Yorktown Heights research lab, both to exercise the
E-beam tool and to ascertain that no unforeseen phenomena
would stand in the way of making such small devices. The news
is quite good. The devices work, and at those dimensions they
could be used to produce the memory chips with about a billion
bits each that I promised earlier in this chapter.

### The Allure of Superconductivity

We had some spectacular news in the late 1980s. Figure 7 shows
the transition to zero resistance when a novel ceramic material
made of yttrium, barium, copper, and oxygen is cooled to a
"critical" temperature of about 90 degrees Kelvin (°K), or 90
centigrade degrees above absolute zero. The discovery of this
fascinating class of high-critical-temperature (high-$T_c$) ceramic

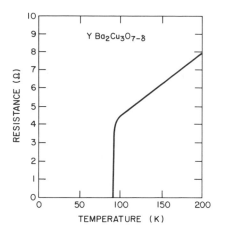

**Figure 7**
The transition to zero resistance in the neighborhood of 90°K for an
yttrium-barium-copper oxide ceramic superconductor.

materials was made at IBM's Research Laboratory in Zurich by
Johannes Bednorz and Karl Alex Mueller, who received the
1987 Nobel Prize in physics for this work.

The discovery of the high-$T_c$ materials took the physics community by storm. In 1987, at the American Physical Society
meeting in New York, thousands of physicists tried to cram into
a room that could comfortably accommodate only a few
hundred in order to hear the latest discoveries about high-$T_c$
materials. And they stayed on until the wee hours of the morning, long after the planned hour of adjournment.

The reason physicists got so terribly excited by the discovery
is that for years they had been frustrated that superconductivity
was achievable only at very low temperatures, requiring cooling
with liquid helium (which vaporizes at about 4°K). They had
dreamed of the day when room-temperature superconductors
might become available. Now, 90°K is well above the temperature (77°K) at which liquid nitrogen vaporizes—and to a
physicist liquid nitrogen's temperature is virtually room temperature. (Liquid nitrogen is cheaper than milk, and is a lot
easier to work with, and more plentiful, than liquid helium.)

So "room-temperature" superconductors were now at hand.
One might ask why we are not yet using them. The answer lies
in a multitude of technical difficulties which must be circum-
vented before these materials can find their way into practical
applications. We have to learn how to reliably make stable

batches of these materials, how to shape them into useful configurations, and how to control their physical properties. Ideally, we would also like to understand them, although we might
be willing to forgo perfect understanding in exchange for adequate control. We are nowhere near that point yet, but we
have been making substantial progress over the past few years.
We now have a pretty good understanding of the structure of
these materials and their dominant characteristics. Composed
of sheets of copper oxide separated by other elements, such as
yttrium and barium, they typically have the "perovskite" structure shown in figure 8. The superconducting property is displayed strongly in the direction of the copper oxide sheets.
Another vexing problem, the fact that the original samples
could pass only a small current before they lost their superconducting property, is not surprising. After all, these are ceramic
materials, like toilet bowls. One does not generally expect
conductivity in such materials. By now we have achieved, in
carefully fabricated thin films, critical currents for superconductivity which are a few orders of magnitude higher than
those obtained in the original samples.

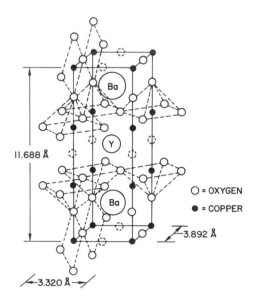

**Figure 8**
"Perovskite" structure, typical of the new high-temperature ceramic
superconductors.

It is still difficult to predict what role these high-$T_c$ superconducting materials may play in the fabrication of computer microcircuitry. The dominant scenario is that they may first be used in packaging for the interconnection of silicon chips, because the current requirements for such interconnections are not as demanding as those for interconnecting devices on chips. Eventually, we may find a way to use them on chips as well. Any such developments would increase speeds, reduce heat dissipation, and be another leap in microcircuitry. This is an exciting prospect, and to say that it is drawing a great deal of attention is a distinct understatement. In fact, it has spurred one of the hottest international races in the computer industry.

## *The Search for Alternatives*

While achieving the exponential improvement of virtually every aspect of computing by pushing an in-place technology to its limits, we have been looking for alternative technologies that will put us on a new path of progress once the old technology reaches its limits. Computers were first made with vacuum tubes, which were replaced by discrete transistors and then by integrated circuits of increasing complexity. Had we stayed with vacuum tubes, one of our big computers of today would be the size of the Houston Astrodome and would require the Gulf of Mexico to cool it. What is worse, it would be unusable, because the mean time between failures would be shorter than the machine cycle.

Today's technological mainstay in computing is silicon semiconductor technology in the form of large-scale integration. Though we are always on the lookout for alternatives, we often come up with silicon alternatives to silicon. But is there life beyond silicon?

Answering this question is not easy. Silicon technology as practiced today has proved remarkably robust, and it has already left some of its challengers in the dust. Several years ago there was a great deal of interest in magnetic bubbles, mainly as a strong candidate for replacing disk storage devices but also as a possible inexpensive alternative to semiconductor memories in some applications. Like disks, bubbles would not be volatile, since they did not require an electric field in order to retain stored information. To be sure, they were slower to read than semiconductor memories. They had to be read serially, but they were simpler and denser than semiconductor memories. Overall, bubbles appeared to fill a gap between semiconductor memories and magnetic disks, being somewhere between the two in speed and potential price. So, what happened? Bubbles never got a chance to get on the "learning curve" of decreasing cost with increasing volume of shipments. Instead, semiconductor memories made such tremendous progress in terms of decreased cost, increased density, and decreased power requirements that they stripped bubbles bare of any claims of superiority for all but a few applications. At the same time, magnetic-disk storage improved in speed and density, squeezing bubbles out of contention. Today bubbles are delegated to a few niche applications that require "rugged" devices insensitive to static electricity and other environmental hazards.

Then there is the story of Josephson technology, which challenged silicon technology for high-end applications during the 1970s. Josephson technology was based on the use of superconducting tunneling switches operating near absolute zero and requiring liquid helium for cooling. They were named after Brian Josephson, who shared the 1972 Nobel Prize in physics for discovering the underlying effect of tunneling between two superconducting domains separated by a thin insulating layer known as the tunneling junction. Circuitry made with Josephson junctions appeared to have huge advantages over silicon circuitry—speeds from 10 to 100 times higher and power dissipation at least 100 times lower. However, during the development of Josephson technology, silicon technology made rapid progress and some technological limitations of the Josephson technology became apparent. In the end, Josephson technology essentially lost the race to silicon, at least for high-end computer applications. Today, it has been delegated to special instrumentation applications using ultrasensitive magnetic-field sensors known as SQUIDs (superconducting quantum interference devices).

These two stories teach us a very clear lesson. In order for a new technology to challenge the supremacy of an established one, it must offer a substantial advantage in performance and a reasonable degree of extendability beyond the capability of the reigning technology.

Are there any viable alternatives to silicon? Is there a need for alternatives? With respect to need, the answer is a qualified yes. At any time, there may be some applications for which the best available silicon circuitry may be inadequate, expensive, or otherwise unsuitable. Also, someday we may reach the point of diminishing returns in exploiting silicon. The qualification is that we never know how good silicon can be. Once more, an alternative technology will be shooting at a moving target. With respect to viable alternatives, the answer is also a qualified yes. A family of compound materials known as *III-V semiconductors,* made up of elements in the third and fifth columns of the periodic table, have some distinct advantages over silicon. Electrons move faster through the lattice structure of those semiconductors, and the properties of layered structures of various types of compounds can be engineered almost at will.

The most popular and the best-understood of the III-V semiconductors is gallium arsenide (GaAs), a compound made up of gallium and arsenic. It is "faster" than silicon, but it is also considerably more difficult to handle. Even the quality of available GaAs crystals is substantially worse than that of silicon crystals, simply because the growth of high-quality compound crystals requires delicate control of the stoichiometric balance (the relative abundances of the constituent elements). Work on GaAs technology has led to the development of MESFETs (metal semiconductor FETs), which have already reached commercialization, and also to the exploratory development of HEMTs (high-electron-mobility transistors), which promise even better performance than the MESFETs.

Silicon is here to stay as the dominant material for computer microcircuitry for many years to come, and for the entire range of computer hardware. GaAs is most likely to be used for niche applications, where the performance of the best available silicon devices is marginal and raw speed is paramount. An example of such an application is the circuitry required at both ends of a fiber-optic cable used in high-speed interprocessor communication links. Such links are used today, but at speeds well below 1 gigabit ($10^9$ bits) per second. They are expected to operate well above the gigabit range as we move further into the 1990s. At those speeds, sending, switching, and receiving signals demand circuitry performance that may be unavailable in silicon. GaAs may then be an inevitable alternative. GaAs also happens to be a material from which lasers and photodetectors are made, a property which is not shared by silicon. This means that using GaAs for the digital circuitry of interprocessor communication links opens up the possibility of *optoelectronic integration*—the building of the circuitry and the lasers or detectors on a single GaAs chip. Optoelectronic integration involving silicon circuitry and GaAs lasers and detectors is also possible if we can deposit domains of GaAs on a silicon chip. This is somewhat difficult because of a mismatch of the lattice dimensions of silicon and GaAs, but progress is being made to accomplish even that feat. This is just another fascinating contest between two competing technologies. Whichever wins, we are likely to benefit from the contest.

It is also very likely that GaAs circuitry will eventually be used to make some computers. Cray Research has already made the commitment to build its next supercomputer with GaAs circuitry, but it has had no followers so far. The number of GaAs computers, and the range of their performance, will depend on our ability to improve the manufacturing processes for GaAs circuitry and to decrease its cost. However, it is not likely that GaAs will replace silicon on a large scale in the near future. Sometimes I think that the Almighty was trying to give us a message when He gave us so much sand. Sand is mostly silicon, and it is there for us to use. It is plentiful, it is not controlled by an international cartel, and it has marvelous engineering properties. We can make almost perfect crystals of it, we can easily make an insulating layer by oxidizing it, and in general we have engineered it so successfully that nothing is likely to challenge its supremacy soon. *Unless,* of course, we make an unexpected discovery. And that's what makes life interesting.

The holy grail of alternatives to current transistor technology, however, is the *ballistic transistor,* a transistor in which electrons fly at ultrahigh speeds. The concept of the ballistic transistor is fairly simple; it is the implementation that is difficult. The concept consists of removing from the path of electrons all the obstacles that normally slow them down as they travel from their originating point to the destination point in a transistor—say, from **Source** to **Drain**. Normally, the electrons are slowed down by collisions with the impurities which are implanted to make semiconductors, and with phonons corresponding to thermal vibrations of the atoms. If such interference could be removed, the electrons moving under the influence of an electric field would continue to accelerate until they reached a very high maximum speed—the ballistic speed.

Collisions slow them down to a small fraction of that speed. Ballistic transistors would give us a quantum jump in speed of microcircuitry, and this makes their pursuit a most worthwhile endeavor. So far, we have achieved ballistic transport of electrons in very thin layers of III-V compounds built with extremely high precision using molecular-beam epitaxy (MBE). We have even achieved a similar feat with *holes,* which are positive charges corresponding to the absence of an electron. These laboratory successes, exciting as they may be, are only the first steps toward the construction of a ballistic transistor, the eventual achievement of which is still far from certain.

### Systems—Large and Small

How do we translate this beautiful hardware into computer systems? This is where *computer architecture* comes in. At first blush, this field does not seem as likely to create a great deal of popular excitement. But it is full of action these days, for at least three reasons: First, advances in internal machine organization and in multiprocessing architecture are contributing to increases in system performance beyond what can be delivered by increases in raw speed of circuitry. Second, personal computers and even more powerful engineering/scientific workstations are invading more and more territories of applications. Third, our ability to interconnect computers in various ways to achieve new functionality has been advancing rapidly.

Consider what is happening to the very large systems used for large business applications and for scientific computation. Advances in circuitry steadily reduce the basic operating unit of time of these machines, the *machine cycle.* During the 1990s, the cycle will be measured in single numbers (rather than tens) of nanoseconds, pushing toward the enchanted goal of the one-nanosecond-cycle machine. Fast as this progress is, it does not give us the performance improvement we would like to have. There is another way of getting performance improvement from a machine: reducing the number of cycles needed to execute an average instruction. This can be done (without changing the machine architecture, which is associated with a sizable investment in software) by increasing the concurrency of computational events taking place inside the machine. There are a host of things one can do in this area—have multiple execution boxes inside the central processing unit, execute instructions out of sequence and merge the results appropriately, and so forth. Such a reorganization of machines should additionally improve performance by at least a factor of 2. Putting all these improvements together, we can expect to increase performance over that of today's machines by something like a factor of 15 or 20 by the end of this century. Not long ago some experts felt that 50 MIPS was an insurmountable barrier to machine performance. Now the 100-MIPS machine is around the corner, and the bar is being raised to 200 MIPS.

We have so far been examining the performance of a single uniprocessor. However, most major installations have, for some time, been using multiprocessing arrangements of various types to get more performance concentrated in a single place. Today's multiprocessing arrangements are *tightly coupled*. They are like the Air Force's Thunderbirds, flying in tight formation—there is a limit to how many processors you can run this way before you start paying a high price for overhead and risking a system crash. We would like to run many more computers together, but like a convoy rather than a tight formation. Thus, we are designing new multiprocessing arrangements which are only *strongly* coupled. They consist of arrays of processors (figure 9) connected with high-speed links to a large data memory and then to input/output subsystems, and they are similarly connected to a message memory and to communication subsystems. In this fashion, we can expect to string together tens of processors to achieve very high performance for a single installation, measured in many hundreds of MIPS. This will also improve fault tolerance, because we will be able to remove any one of the processors from the picture without crashing the system. The challenge, of course, is to make such a multiprocessing arrangement "upward compatible" with today's systems, so that it can make use of the large investment in existing software. It is not very surprising that this is largely a software challenge; in almost anything we do today to improve systems, the software challenge often outweighs the hardware challenge.

What about the small systems, ranging from the inexpensive personal computers to the sophisticated engineering/scientific workstations? The computational power of these systems appears to be increasing even faster than that of the large systems. It is achieved by cramming more and more FET transistors

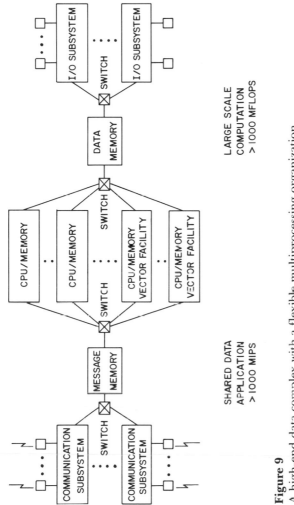

**Figure 9**

A high-end data complex with a flexible multiprocessing organization allowing expansion to a parallel operation of tens of processors, either for commercial applications or for large-scale computation.

onto a single microprocessor chip. There is better than a linear relationship between microprocessor performance and the number of transistors on a chip. In the late 1980s, the most popular microprocessor chips had up to about 300,000 transistors and a performance of about 5 MIPS. Some chips with over a million transistors appeared by 1990, supplying something like 20 MIPS. If we were to plot the projected performance of microsystems together with that of large mainframe uniprocessors, we would probably see the two lines intersecting sometime around the middle of the 1990s. This has caused some people, particularly those who happen to be strong proponents of small systems, to predict the impending demise of large systems. I consider such a demise unlikely, although the allocation of investment in large versus small systems will undoubtedly change in the coming years. Large systems will continue to be used to maintain large databases, to provide a rich menu of services, and in general to provide the engines for the large transaction-oriented systems which influence so many aspects of life today.

There is an analogy between the range of computer sizes and the range of transportation vehicles that we use. Personal computers are like personal automobiles. There is a limit to the price which most of us are willing to pay, but on the other hand we do not insist on amortizing the cost through full use. The personal car is valued for the convenience of ready availability and point-to-point transportation. Public transportation, on the other hand, involves a greater investment in capital and labor, which is amortized through substantially greater utilization. The reliability and availability requirements are also higher for public transportation. But the ease of use is more of an issue for private cars. Finally, private cars and public conveyances coexist, and are likely to continue to coexist in the future, although the allocation of investment between the two changes from year to year. Similarly, in computers, it is unlikely that small systems will displace large ones, particularly since to a small system a large one is just some other large peripheral device which provides it with convenient services. The rate of investment will continue to change, as it has been changing over the past several years. Whether or not the internals of large and small systems will use similar microcircuitry will depend upon the evolution of the various flavors of circuitry, and upon its use both in well-established computer architectures and in some new ones which have appeared during the past few years.

### RISC versus CISC

There is another interesting debate underway: RISC versus CISC. RISC, the reduced instruction set computer, is the brain-child of John Cocke, an IBM Fellow, whose genius I have come to respect more than that of anybody else I know in the computing world. In 1975 Cocke asked a very important question: "How would one design a computer without any architectural constraints, using the hardware that we have available today and all the tools of the trade that we have acquired over the years?" He and his colleagues sat down and analyzed "trace tapes" of computer runs, and found out that computers performed relatively simple tasks most of the time, whereas some of the tasks embodied in complex instructions were executed less than 10 percent of the time. They decided to build hardware that would execute only the most useful computing tasks, but would do so extremely fast. Thus was born the concept of the IBM 801 minicomputer, named after the IBM Research building at Yorktown (which is Building 801 in the IBM real estate inventory). The 801 prototype, completed around 1980, revolutionized the thinking of computer architects around the world. It also spawned a raging controversy of RISC versus CISC (complex instruction set computers). The proponents of CISC argue that RISC has taken us back to the days when computers had only simple instructions. They say that RISC is throwing away all the progress that has enabled us to *microcode* (build in hardware) beautiful, complex instructions which can execute very complex tasks but which require fetching only one instruction from memory. And they assert that making the instructions simple holds us captive to the infamous "von Neumann bottleneck," the congested link that brings instructions from memory to the CPU. Obviously, their arguments fall on deaf ears, because the industry has adopted the RISC concept with great enthusiasm. Every major computer manufacturer, and quite a few minor ones, have RISC products out or under development. This questions the merits of the controversy.

Here is a facetious attempt to emulate Werner Heisenberg in resolving the RISC-versus-CISC controversy. If we defined the complexity of the instructions as $C$ and the speed of execution as $S$, then the product of these two quantities, perhaps with a power correction for one of them, might be a constant, $g$ (the reason for choosing $g$ is unknown): $CS^k = g$.

The CISC computers lie in the range of high complexity but relatively low speed, and the RISC computers lie in the range of low complexity and high speed. I cannot resist defining the limit designs at both ends of the spectrum: the ICISC (infinitely complex instruction set computer), which has infinitely complex instructions but executes them at zero speed, and the NISC (null instruction set computer), which has a null set of instructions but executes it at infinite speed.

It is unlikely that the CISC computers, which evolved with all the baggage they had to carry for such reasons as software compatibility, could have hit the optimum in instruction complexity. John Cocke was right in asking the question he asked. Moreover, progress in technology has given us new capabilities over the years. The balance between the speed of logic circuitry and the access time of memories was such that it made the RISC machines advantageous. And the reduced instruction set required less silicon "real estate," which gave some additional benefits. For example, John Cocke and his colleagues used some of that real estate to supply many registers and use many register-to-register operations, eliminating many of the necessary journeys into memory. Also, by creating a "lean and mean" set of simple and uniform-length instructions, they made possible the construction of a highly efficient *optimizing compiler,* which permits the programming of new instructions, and all programs (including systems programs), in a high-level language. This made a substantial contribution to the solution of the software-development bottleneck. CISC architectures will continue to be used, if only for reasons of software compatibility. But, all things considered, RISC has earned a prominent place in the pantheon of architectural concepts, and is already on an evolutionary path to even greater performance. The most recent RISC machines use such innovations as a *superscalar* architecture, capable of executing more than one instruction in one machine cycle. It is a special form of parallelism involving a few specialized processors, including a highly efficient *floating-point* processor, to be used for numerically intensive computations. They achieve performance measured in tens of MIPS at workstation prices. Clearly, RISC architectures have established for themselves a place in the pantheon of computing.

### Connectivity Unlimited

Every computer now wants to get on the phone and talk to another computer. Computer networking has already been around for many years, but the issue of connectivity is now pivotal. In the very near future, we will no longer be satisfied with either the range or the speed of the connections that are available to us today. We will expect high-speed links, and we will expect them to be available for every source of information that we may need. Most likely, we will expect to have access to unused resources located elsewhere in the most transparent way, without a lot of "handshaking" and non-value-added communication.

The high-speed links are becoming available with the advent of fiber optics, but they will require a revisiting of the protocols for interprocessor communication. Today's protocols, designed for low-speed telephone interconnections, tolerate a lot of unproductive handshaking, acknowledgements, and so forth. As the speed of transmitting useful information increases, this overhead becomes intolerable. It will have to be reduced through the use of improved high-speed communication protocols.

Another major trend in designing computer networks is to make them less susceptible to interruption than they are today. This can be achieved by decentralized control of their operation, allowing action to correct a local failure (such as a line failure) to be implemented locally and dynamically by the controller of a neighboring node. This would make computer networks dynamically reconfigurable, instead of having them operated according to predesigned strategies of point-to-point interconnections (route tables and so on). A dynamic routing strategy would also make the operation of very large networks much more feasible than is possible under a fixed-route strategy. The latter approach becomes unwieldy as the number of nodes increases, because of the even more rapid increase of the size of the requisite route tables stored at various nodes.

Today, a PC owner uses his computer either as a stand-alone workstation, or as a terminal to access a mainframe over a public branch exchange (PBX) or through some other dedicated link (figure 10). In switching from one mode of operation to the other, the user generally switches "environments" from the MS/DOS operating system environment to that of the main-

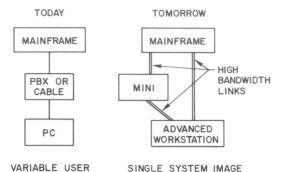

**Figure 10**
Use of personal computers today, contrasted with likely evolution of this use
in the future.

frame operating system. This is fine for those of us who have
been conditioned to put up with the idiosyncrasies of computers. It is fair to say that, over the years, remarkable success
has been achieved in making the users friendly to the systems.
Now, however, the time has come to pay attention to making
systems friendly to users. Thus, future systems will offer a *single
system image,* such as that shown on the right side of figure 10.
Users will have at their disposal ever-more-powerful "super
PCs" capable of doing very complex tasks independently. But,
in addition, by issuing a *virtual system command* they will be able
to access, transparently, additional data and services at the appropriate level of a hierarchy of computers. This will provide
various rich mixes of services, possibly with higher guaranteed
performance than average users can provide for themselves.
Some of the elements of this organization are already in place,
including common operating systems for PCs and mainframes.
What remains is to put all the pieces together in order to
achieve this best of all possible worlds.

### Beyond von Neumann

If we ever decided to establish a Computing Hall of Fame, John
von Neumann would undoubtedly be its first inductee. Ironi-
cally, his name is also associated with the most notorious
scourge of computing, the *von Neumann bottleneck.* This limitation plagues all conventional architectures, which are variants
of von Neumann's original concept of the stored-program ma-

chine. It is caused by the fact that all instructions and data have to be fetched from memory to the CPU and back, one by one, across a link whose traffic-carrying capability limits the machine's performance. The most effective remedy is *parallelism*. The 1980s were a period of fruitful experimentation in parallel machine architectures. The 1990s will bring the fruition of these experiments and make available an unprecedented supply of computing power.

Like everything else in computing, parallelism comes in different flavors. We have already seen a form of parallelism in the multiprocessing arrangements which are being widely used. This is a low degree of parallelism, and somewhat limited in scope. It can accommodate well a stream of relatively small jobs, but it cannot be easily used for a single, large, scientific computation. The next level of parallelism is intermediate parallelism, involving up to a few hundred processors which function essentially as a uniprocessor. Beyond that are massively parallel architectures involving thousands of processors. Several experimental computers of the intermediate and massive levels of parallelism have been built, and they are setting the stage for rapid development of this technology during the 1990s. Most of the successes in this area have involved what one might call the construction of a special-purpose computer. We have, in effect, an "existence proof" that, given a specific task, we can build a parallel architecture that handles it well. What we have not achieved with equal success is the construction of a general-purpose (or at least a multi-purpose) computer. This is, to a great extent, due to the fact that a general-purpose computer should not just comprise a hardware design that works effectively for various types of tasks. It should also come with a software package, akin to a compiler, that makes possible the programming of those tasks for execution in this computer.

Figure 11 illustrates two basic generic configurations which are prime architectural candidates for intermediate parallelism. In one of them, many processors, each with its own memory, are connected through a fast switch; in the other, many processors are connected through a fast switch to a shared memory, which can be a collection of memory boxes. These concepts have been combined by IBM Research and New York University in an experimental multi-purpose computer known as the RP3 (research parallel processor prototype). Up to 512 proces-

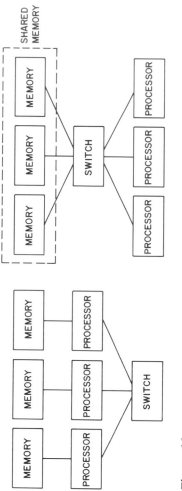

**Figure 11**
The two main generic classes of parallel architectures. In the first, many computers, each with its own memory, are connected over a fast switch. In the second, many computers are connected over the switch to many memory units or a common shared memory.

sors, each with a processing capability of about 2 MIPS, may be connected to form a complex capable of delivering about 1000 MIPS. Each of the processors has a memory which is partially local and partially global, accessed by the other processors through a very fast switching network. What distinguishes this computer from the well-publicized supercomputers already in use today is that its high performance is not limited to highly vectorizable tasks (tasks which have a high content of computations involving identical operations on large arrays of numbers). It is equally good for tasks with any vectorizable content, provided that they can be programmed to utilize the available computing resources effectively. To make it useful, this machine comes with software permitting relatively easy programming of a variety of tasks.

The RP3 is intended as a laboratory of parallel architectures, and as an existence proof that a general-purpose parallel machine can be built. I fully expect that commercialization of multi-purpose parallel machines will flourish during the 1990s.

Experimentation with massive parallelism is lively but still in a state of flux. One of the earliest entrants in this area was MIT's Connection Machine, in effect a very large intelligent memory connected to 16,000 processors. It is being used to tackle various tasks, such as image processing, that are particularly suitable for execution in such a machine. Other current or potential massively parallel projects involve the use of a notable British entry in the RISC and parallel-architecture fields: the Transputer, built by INMOS, Ltd., which is intended as a building block for parallel systems. I expect that the Transputer will leave its mark in parallel systems in the coming years. The extent of the mark will depend on the evolution of the Transputer as a flexible building block for architectures yet to be invented. Unquestionably, the Transputer with its built-in interconnection gear, its low cost, and its special programming language for parallel execution (OCCAM), is providing a convenient platform for experimentation in parallel systems.

Just over the horizon loom the *neural networks,* which combine concepts of parallel architecture with those of artificial intelligence and machine learning. They are the current darlings of the popular and trade press, just as "expert systems" were a few years ago. They are based on a concept that relies on our understanding of how a human brain learns to recognize patterns by establishing interconnections of its neurons. A computer version of a neural network consists of many processing elements, with an initially indiscriminate but rich interconnection which is continually modified through learning. As the network is exposed to inputs, such as those produced by viewing an image, certain links are strengthened and can later respond to the same pattern more readily than to a random pattern.

Like everything new that comes up in computing, neural networks were oversold as a solution to most of our problems. This "hype" stage is generally succeeded by a mature realization of the limitations of a concept and the hard work that must be done before the concept can be made useful. Finally, the applications that survive scrutiny and the test of time are commercialized. Neural networks are now approaching the middle stage of their development, having survived the early stage.

Some very useful and productive theoretical work on *self-organizing synaptic networks* is slowly providing a scientific underpinning for neural-network activities. The extent of the eventual success of neural networks will depend on how many useful applications can be mapped into commercially viable configurations. Neural networks seem to be prime candidates for niche applications involving some form of pattern recognition—applications which are not handled very efficiently by conventional computer architectures.

### User Interfaces

Here is the best news. All those MIPS which we seem to be able to produce at ever-lower prices will continue to make living with computers easier and easier. We have already seen this phenomenon in the way we use PCs. Most of their processing power is used to present a convenient interface to the user, be it a word-processing environment, a spreadsheet interface, or whatever. The trend will continue, with improved quality of presentation and improved modes of computer interaction.

Today's input/output devices operate at data rates which are generally under 1000 bits/sec, and are driven by processing power well under 1 MIPS. As the available processing power increases, and as we learn how to make use of parallelism, we will be able to use processing power of 1000 MIPS or more and achieve data rates 1,000,000 bits/sec and above. Increasingly, the input devices that we use today, such as keyboards and touch-sensitive screens, will give way to more convenient modes of interaction with computers, through natural-language input, speech, and handwriting. Moreover, the quality of presentation will be a far cry from the current grainy, jagged, static images which most users get on their screens today.

### Visualization and the Evolution of Computer Graphics

There was a time when computer graphics was a luxury item. Now it is a necessity. Computers produce such vast amounts of information that it is often impossible for us to digest them. This is particularly true in scientific/engineering applications, where supercomputers are producing output numbers faster than anyone can absorb them. And so *visualization* has become a code word. It describes the conversion of numerical data into screen images, still or moving, which can present us with all useful information, with near photographic quality, in color. Visualization has become a legitimate discipline and research topic, and another essential tool in computing. Figure 12 shows a visualization of the instability of shear flow in a viscous liquid, which causes the formation of vortices reminiscent of the flow in the wake of a body moving on the surface of a liquid. This visualization of turbulence was obtained by solving a set of hydrodynamic equations on a supercomputer and transforming the reams of output paper into convenient images on a color display.

Architects and other creative artists can use computer visualization to produce realistic representations of objects before they are built. They can (for example) generate a faithful representation of the view of the landscape from inside a virtual building, and experiment with the design until they get it just right. To get the effects of illumination, reflections, and shading of complex objects takes a lot of computing power. But we are getting more and more of this power every day.

Here is another example of the power of visualization. Starting in the 1960s, Benoit Mandelbrot, an IBM Fellow, developed a beautiful new mathematical concept, the concept of *fractals*. Fractals are objects that are not one-, two-, or three-dimensional, as are most objects with which we are familiar. The dimensionality of fractals is fractional, between the integers 1 and 3. The concept describes the fact that certain things in life display structure no matter how much we "zoom in" on them with

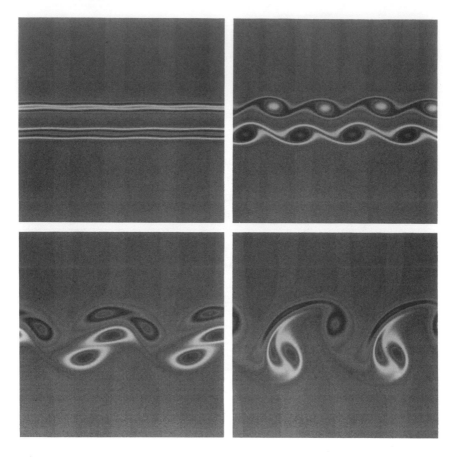

**Figure 12**
Visualization of turbulent fluid flow.

higher and higher magnification. Mandelbrot gave an early illustration of the concept by pointing out that the length of the
coast of England (say) changes as we change the scale of maps
and display a finer and finer structure of the coastline. The
dimensionality of the coastline ends up being somewhere between that of a line (1) and that of a plane (2). Mandelbrot
investigated fractals and showed how we could create our own
versions of such objects, providing a most useful emulation of
the process by which nature generates such seemingly random
shapes as coastlines and mountain ranges. Fractal geometry fascinated scientists, who used it to describe a plethora of natural
objects. But fractals have also shown up in landscapes in the
film *Star Wars* and its sequels. These dramatic scenes can be

**Figure 13**
A planet and a moon-like landscape generated on a computer using the
fractal geometry of Benoit Mandelbrot.

traced to the beautiful visualization of fractals produced by
Mandelbrot's collaborators, including Sigmund Handelman
and Richard Voss. Figure 13 exemplifies Voss' renditions of
fractal landscapes and planets, which gave birth to this new
form of art.

The future looks very bright for computer visualization and
related technologies. We already have, in computer animation,
almost a new medium of entertainment. Further down the line,
computer processing of images, real or man-made, is going to
be an everyday occurrence in every household. The distinction
between a computer and a television set will slowly disappear.
The future TV set will have substantial capability for processing digital images received over any of several available communication links. Abundant bandwidth will make possible
*narrowcasting*, giving viewers a wide choice of entertainment or
other offerings. Furthermore, the viewers will be able to edit
and direct the playback according to their wishes. They may
even be able to edit the faces of actors. All these developments
will take place before 2001. It has been said that the current
debates about high-definition television (HDTV) are moot, be-

cause HDTV already exists in the computer domain. All that remains is to finalize the blending of the domains of computing and television, and the associated standards.

### Natural-Language Processing

Researchers at the Yorktown lab of IBM Research have made two sorties into the processing of natural language. One is a program that can critique a text that has been entered into a computer, identifying grammatical and syntactical errors, the use of wrong homonyms, and even stylistic imperfections such as repeated use of the same adjective. Then, it provides competent help toward improving the text. The other of these sorties is an interface to a relational database query facility that permits a query to be made in natural English. The program extracts the meaning of the query, perhaps ascertains this meaning interactively, and then translates the query into an artificial language acceptable as input to the database program. The first program has been in experimental use for a few years, with very favorable results. The second program has been used, experimentally, to query a database of real estate parcels. At this stage, there is no question in anyone's mind that such applications involving input of natural language are eminently possible. The only questions are how acceptable these applications are to the users and how much they cost. As we get closer and closer to the (possibly naive) end user, human-factors considerations are the determinants of the success or failure of a particular application. Natural-language processing will happen, but its degree of pervasiveness will depend on its user friendliness.

### Speech Recognition

Speech is the most natural form of human communication. And speech as a form of computer input has been pursued by many over the years. We have made considerable progress. Today, one can buy inexpensive PC cards, capable of recognizing a limited vocabulary, which can be used for applications involving speech input. However, the task of recognizing unconstrained speech, particularly continuous speech involving a large vocabulary, is immensely more difficult and still largely elusive. Considerable progress has been made in this area since 1970 by Fred Jelinek and his colleagues at IBM Research.

They have tackled the problem not by trying to emulate the cognition capability of a human being, but by using the computer to do things *it* can do well. They have applied concepts of information theory to the task of speech recognition. They taught the computer the statistics of language—the fact that a segment of speech does not appear randomly in nature but depends on what precedes it. They used such statistical information to bootstrap the acoustic pattern recognition. And finally, by imposing a slight constraint on the speech input, a slight pause between successive words, Jelinek and his collaborators managed to build a prototype which has impressive speech-recognition capability. It can recognize a vocabulary of 20,000 words with an accuracy of at least 97 percent, and it does this in real time (permitting immediate correction of mistakes). It uses nothing more than an IBM PC/AT equipped with some special cards that use a special, fast signal-processing chip. We are poised at the threshold of being able to add "listening typewriters" to the tools spawned by computer technology. Once more, there is no doubt that we can build something; the question is how cheaply, and how acceptable the speaking environment is to the potential users. The next few years will give us the answers. I certainly think that we will be using some versions of sophisticated speech input during the 1990s.

### Expert Systems

The idea of expert systems is an idea whose time has come. The idea is to capture the expertise of a human being, translate it into a convenient form usable through a computer, and make it available to another person in an interactive consultation mode. It has already been demonstrated successfully by implementing consultation programs in medicine, oil exploration, and computer system diagnostics.

Expert systems started out being somewhat oversold, and this may have created a backlash among early users. But now we know pretty much where expert systems can be used successfully, and their development is moving along at a brisk pace. We are building larger and larger expert systems, and we are creating the tools for building more of them conveniently (e.g., an understanding of the knowledge-acquisition process, languages for programming the inference-oriented programs used in expert systems, and *shells* that can be filled with knowledge). Expert systems have thus become a modern, well-organized version of decision support systems, and of assistance programs for carrying out diverse activities (including such challenging tasks as software development).

There are things in life that probably will not be translated into expert systems. Certain things that humans do by intuition cannot be captured and organized into such a well-disciplined form. There are other things which we cannot do very well with expert systems today, but we will probably be able to do tomorrow. My favorite example comes from the fact that we cannot conveniently merge conflicting expertise of different experts into an expert system. Now, human beings have the uncanny ability to combine individual ignorance, by means of a task force, to produce collective wisdom. We cannot make task forces out of computers today. But perhaps some day we will do even that.

# 2

# *Ordering Chaos: Supercomputing at the Edge*

*Kathleen C. Bernard*

. . . any country which seeks to control its future must effectively exploit Very High Performance (VHP) Computing.[1] A country aspiring to military leadership must dominate if not control VHP computing. A country seeking economic strength in the information age must lead in the development and application of VHP computing in industry and research.

*White House Science Council Committee on Research in Very High Performance Computing*

Supercomputing—solving problems at the leading edge of science and engineering—will do much to determine the rate and the quality of technological innovations and the rate of economic growth into the new millenium.

Technological innovation will intensify through the 1990s: breakthroughs in the underlying sciences will be decisive. To date, our understanding of physical systems and phenomena rests on collections of systems that were primarily described computationally as linear. As physical systems continue to shrink (e.g. semiconductors), or where they span vast scales of time and space (e.g. global climate change), the description of systems will become more complex, involving multiple science disciplines; nonlinear effects will become more important.

Commercially relevant R&D is inseparable from competitiveness. Innovation includes breakthroughs as well as specific applications. Yet the United States has been slower to incorporate technological innovations into commercial products and services than some of its trading partners.

"Supercomputers" are the most powerful computers at any point in time. The perishability of the definition illustrates the speed with which supercomputing technology is unfolding.

The capability of the supercomputer of the mid-1970s will be available on personal workstations in the 1990s. An early serial number of the pioneering Cray-1, the performance benchmark for high-performance computers, is now in the Smithsonian's Air and Space Museum.

The capability provided by supercomputing is a function of the raw performance embodied in the hardware and of the ability of the system to move and access data at a commensurately high bandwidth—into and out of memory as well as to secondary storage. The tight coupling of the data handling with the processor capability has implications for technology directions of the future—especially in the memory and storage arenas.

The architectural feature associated with supercomputers in the mid-1970s was vector processing. Vector processing allowed large groups of numbers, or vectors, to be processed in parallel, resulting in performance speedups by factors of 10 or greater (whereas generational improvements in more traditional computers offer twofold or threefold speedups). In the early 1980s, parallel supercomputing was introduced, allowing multiple processors to work concurrently on a single problem. By the end of the century, significantly greater computing parallelism (combining perhaps tens of thousands of processors working effectively on a single problem), and architectures that integrate multiple modalities (such as hybrid combinations of numeric and symbolic computing) may be possible. Probably through this decade, software problems will restrain the productivity which could otherwise result from parallel supercomputing: computer programs that model tomorrow's complex tasks will be significantly more difficult to write, debug, validate, and deploy than the programs of the 1980s.

Supercomputers enable researchers to model complex phenomena with increasing realism. Complex models can incorporate multiple scientific disciplines, greater accuracy, and complicated relationships among variables and subsystems. For example, chemical, physical, and metallurgical processes might be coupled to fluid-dynamical equations in order to illuminate chemical vapor deposition (a process important to the semiconductor industry.[2] Problems can range widely in scale—from nanoseconds to years, and from macrostructures to microstructures. By combining multiple scientific disciplines and scales into a single problem space, supercomputers allow detailed investigations at the edge of what had been impossible achieving greater understanding through numerical experimentation of what had been theoretically described with equations.

Supercomputers open up new realms for investigation and enable greater problem domains to be considered. Researchers can develop solutions that treat entire problems from first principles—building from interactions at the atomic or the molecular scale. As the tool of researcher imagination, new insights are possible and new approaches to problem solving are discovered. The researcher is constrained only by imagination.

### *The Applications*

George Stalk, Jr., notes that "the ways leading companies manage time—in production, in new product development and introduction, in sales and distribution—represent the most powerful new sources of competitive advantage. . . . in fact, as a strategic weapon, time is the equivalent of money, productivity, quality, and even innovation."[3] Relative time is what supercomputing buys. Projects dependent on computationally based R&D can be completed in shorter periods of time. More alternatives can be evaluated, pointing the way to better designs, more cost-effective products, and more efficient implementations. Failure can be computationally anticipated; higher researcher productivity is realized. Technological innovation is increasingly dependent on science-based R&D. Attempts to advance the state of the art can actually be limited by the current level of knowledge of the underpinning disciplines. The increasing use of computational science as an intermediary between theory and experimentation can help reduce these limitations.

Steven Klin and Nathan Rosenberg write:

The development of new alloys with specific combinations of properties proceeds very slowly because there is still no good theoretical basis for predicting the behavior of new combinations of materials; the same applies to pharmaceutical drugs. Many problems connected with improved pollution control are severely constrained by the limited scientific understanding of the combustion process, and by the fact that the design of a combustion "firebox" remains in 1985 still an art based primarily on the results of prior designs—not on science. The development of synthetic fuels is at present seriously hampered by scientific ignorance with respect to the details of oxidation reac-

tions in various forms of coal. The designs of aircraft and steam tur-
bines are both hampered by the lack of a good theory of turbulence.
In the case of aircraft, wind-tunnel tests are still subject to substantial
margins of error in terms of predicting actual flight performance.
Indeed, in considerable part the high development costs for aircraft
are due precisely to the inability to draw more heavily on a predictive
science in determining the performance of specific new designs or
materials. If science provided a better predictive basis for directly
specifying optimal design configurations, development costs (which
constitute about two-thirds of total R&D expenditures in the US)
would not be nearly so high.[4]

Research is performed to solve problems in all elements of
the product-development process—from design through manufacturing. Although supercomputers have been used mainly
in pure scientific research, this is changing. Today they are
being used to design products and to improve the processes by
which products are developed. Design is becoming a strategic
activity, including manufacturing and other phases of the product's life cycle. Increasingly, product design is the result of analyzing many design possibilities, resulting in more efficient
product development. Some studies have noted that perhaps
70 or 80 percent of the ultimate cost of a product is determined
by the end of the design phase.

There are many examples of the use of computational power
to address economically important problems of industry.

Supercomputers have been applied extensively in *structural
analysis* and in analyzing the *dynamics of fluid flows* about large
structures (for example, ships and aircraft). The speed of a supercomputer permits rapid analysis of many alternatives before
a commitment must be made to build a prototype to a particular design. More thorough evaluations of different aircraft
body configurations have resulted in improved aircraft range
and fuel efficiency. Research in combustion has also accelerated
these solutions.

In *combustion research* dedicated to greater engine efficiency
and less pollution, supercomputer estimates of such variables
as chemical concentrations throughout the combustion chamber offer an immediate understanding of how a given design
performs.

*Automobile manufacturers* increasingly compete by delivering
qualitative features such as "quiet rides" or "climate control."
They apply advanced computational fluid dynamics not only to

modeling exterior air flow, as is done for airplanes, but also to simulating the flow of air in the interior compartments. Before supercomputers, analyzing interior air flows required a working prototype. Problems in air flow discovered at such a late phase were extremely expensive to correct. Supercomputers have also led to entirely new processes in industry. For example, crash analysis, where computer simulated crashes replace crashes with actual cars, is one of the fastest-growing supercomputer applications in the automotive industry.[5]

Supercomputers can be used in *materials design*,[6] allowing complex phenomena to be described realistically. For example, improving the manufacturability of certain lightweight structural alloys is valuable to the highly competitive aircraft industry. Developing and processing lightweight yet high-strength structural alloys is of great importance, but manufacturability is constrained by a material's ability to be shaped and by its propensity to develop cracks and fractures. Understanding of cracks and fractures is enabled through detailed supercomputer models of the microstructure.

Another major manufacturing activity that could be improved by supercomputer analysis is *metal forming*, which costs the auto industry billions of dollars per year[7] and which currently involves a great deal of trial and error. Forming sheet metal into parts requires precisely shaped dies and careful specification of the steps necessary to define the ultimate shape.

Supercomputers also make possible the evaluation of *environments where physical experimentation is not feasible,* as in the nuclear power industry.

Supercomputers provide an overarching view of an entire system and allow simultaneous evaluation from the micro scale to the macro scale. The macro scale enables us to observe long-term effects (on the order of decades); on the micro scale, we can model the behavior of a system at the molecular or atomic level. This makes it possible to understand the dynamics of completely separate physical systems—such as the interactions of the atmosphere, the physical and organic structures of the environment, and various pollutants. An example of the need to incorporate ranges of scales is the US government's thrust for advanced R&D in analyzing "global climate change."

In the past, the US government has purchased the world's most powerful supercomputers. Government researchers have described their most computationally intense problems as "grand challenges." Examples of these "grand challenges" include the following[8]:

predicting weather and studying global climate change

designing new materials, and analyzing atomic structure for that purpose

understanding how drugs affect the body, and designing new drugs

understanding the interplay of the flows of various substances involved in combustion systems and the quantum chemistry causing those substances to react

locating oil, and devising economic ways of extracting it

modeling the fluid dynamics of three-dimensional flow fields around the shapes of complete aircraft.

Most of these "grand challenges" are considered to be beyond the current capabilities of supercomputing, depending on the problem size and the degree of complexity the model entails. Problems such as these can require a trillion floating-point operations per second, and billions of words of memory. This is a thousandfold greater requirement than the capabilities offered in 1990.

In addition to addressing the technological challenges of competitiveness and offering a better understanding of complex systems, supercomputers contribute to national security. Supercomputing is viewed as a strategic capability that can definitively affect the balance of power in the world.

### Leveraging R&D

Numerical experiments make productive interaction between physical experiment and theory possible. Using numerical calculations based on theoretical concepts, a physical system is modeled, and its behavior is predicted. Numerical experiments involve developing computational models and coding them into computer programs, known as *application codes*. Previously unknown phenomena are discovered. Often, data are available that would not have been possible through physical experiments, where the scale of events can be beyond our ability to detect them. Variations from what we would expect cause us to reevaluate either the theory or the experimental methods, or both. For example, in some cases of combustion research, disagreements between experiment and computational simulation have revealed errors in the experiment approximately half of the time, with the remaining errors due to deficiencies in theory.

The complexity associated with a problem is a function of the algorithms used to solve the problem (operations per variable), the degree to which first principles are reflected in the model (variables per point), the time steps required to solve the problem (steps per problem), and the specific geometry of the problem (points per step). The capability of the computing tool used to solve a problem determines the amount of complexity that can be treated—thus, it determines the degree of realism that can be incorporated into the model for evaluation.

Many problems can now be modeled in only two dimensions. While greater complexity enables new problems and scales to be included in an experiment, it also dramatically increases the problem's size. Similarly, increasing a problem's accuracy adds to the number of grid points used to solve the problem. This affects the problem's size, and perhaps even its tractability. Tractability is a function of the required solution time, of the problem's complexity, and of the computing capability to be applied, as depicted in figure 1. Many complex problems can run hundreds of hours. Problems that take this long are usually considered intractable, because they take too long to complete. Typically, such problems must be "run" a few hours each day, as computing resources become available.

Desiring to make daily progress, researchers constrain the science in their numerical simulations to a maximum execution time of about 10 hours. They like to work with their problems interactively, ideally having the opportunity to run two or three "experiments." This then requires a much shorter solution time. Problem complexity is scaled to the computer's ability to produce results in a manageable amount of time. For example, the estimated complexity of large problems being solved on current supercomputers is between $10^{13}$ and $10^{14}$ operations per problem.[9] As can be seen from figure 1, problems requiring $10^{14}$ operations per second can be solved in approximately one day using supercomputers. "Grand challenge" problems, such as global climate change, are even more complex. Many problems are intractable either because the computer power necessary to address the problem in a reasonable time is not available or the sheer size of the problem is too great, or both.

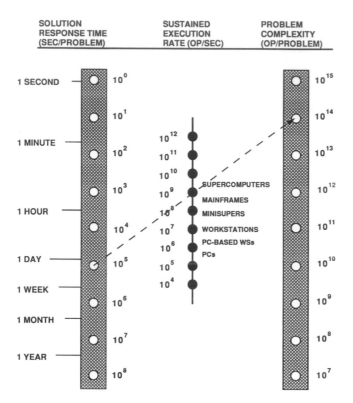

SOLUTION RESPONSE TIME (SEC/PROBLEM)

SUSTAINED EXECUTION RATE (OP/SEC)

PROBLEM COMPLEXITY (OP/PROBLEM)

**Figure 1**

Nomograph of computer simulation. Adapted with permission of Worlton & Associates.

Increasing a problem's complexity can also imply greater data intensity. Typical numerical experiments can produce quantities of data in the range of tens of gigabytes. Often the ability to treat complex problems is limited by the speed at which the computing system moves this data into and out of the system. Relative slowness in moving data, or the sheer volume of data related to describing complex problems or resulting from the calculations of a numerical experiment, can create additional sources of intractability.

Crucial to quickly building on new understandings resulting from numerical experiments is the researcher's visual interpretation of and interaction with the "experiment"—a process known as *visualization*.

Visualization has become indispensable to the supercomputing research environment. It "enriches the process of scientific discovery and fosters profound and unexpected insights; in many fields it is already revolutionizing the way scientists do science."[10] Rather than simply viewing static models of molecules, for example, researchers view molecules actually vibrating and moving about. Visualization enables the researcher to "see," in real time, errors or limitations in the numerical experiment. This allows immediate correction, enhancing the productivity and quality of the experiment while allowing feedback to the theory on which it is based. Researchers can not only interpret what is happening to their data during the "experiment"; they can interact with it almost in real time, changing the parameters, the resolution, or other aspects of the model and seeing the effects. Visualization can itself be a computationally intense problem.

### What Lies Ahead?

Early supercomputing systems of the mid-1970s performed at peak rates of roughly 100 megaflops. By the early 1990s, supercomputers performed at approximately 2000 megaflops. Reasonable forecasts anticipate 100 gigaflops in the mid-1990s, and a teraflop supercomputer appears feasible by the end of the decade.[11] However, theoretical performance targets delivered through the architecture are rarely completely achieved, because of lags in software development. Clearly, the challenge through the end of the decade will be researcher productivity, as researchers address the "grand challenges" of the 1990s. A crucial underpinning of this effort will be the progress of the software necessary to support the rapidly increasing levels of parallelism that will be focused on a single problem.

The need to tackle more complex problems has focused attention on revolutionary approaches to increasing system performance. New architectures incorporate tens of thousands of processors that strive to work together on a single problem. Increasing parallelism will provide the computer performance needed to solve the teraflop science and engineering problems of the future. The emphasis on parallelism is due in part to government "challenges" that seek to realize teraflop performance by the mid-1990s. Other options for achieving such computational performance will be through hybrid architectures, such as ones that combine numeric and symbolic computing.

The second half of the 1990s will see dramatic shifts in the way science and engineering research is conducted. Computational science will take on a larger role. Supercomputers will be able to probe events and scales beyond the reach of physical experimentation. Solutions will be developed by multidisciplinary teams of computer scientists, algorithmists, computational scientists, manufacturers, and private-sector and public-sector researchers.

### Tomorrow's Computational Laboratory

Workstations which deliver performance approaching that of the supercomputers of the early 1970s have become available in the early 1990s. Personal supercomputing is possible from the researcher's desk. Dynamic, lifelike visualization of experiments allows researchers to "zoom" in to micro scale or out to observe macro-scale results of numerical experiments. The researcher's ability to attach physically remote resources to his environment personalizes the laboratory as well.

The numerical laboratory of the future is a fully integrated, distributed computational environment—bringing to fruition the idea of "network supercomputing." The "network supercomputing" environment effectively combines high-power workstations, supercomputers, file servers, artificial intelligence stations, and other unique R&D tools into a personal capability available to individual researchers. The network is sufficiently fast at the supercomputing hubs that the transfer of large input or output data files associated with "grand challenge" problems does not cause perceptible delays for the users. The researcher "sees" all network resources that are public, or to which he has access, as if they were physically under his control—a "virtual" laboratory regardless of location.

The environment is nearly seamless, allowing researchers to share experiments transparently—from the workstation platform to the supercomputer platform to the file server. The operating systems of the various platforms are sufficiently similar that we can speak of "interoperable platforms." This interoperability allows researchers to move easily from one numerical laboratory to another. Alternatively, supercomputing cycles can be "wheeled" around the network, creating a supercomputer-cycle grid analogous to the grid of today's electrical power system. With high-speed interconnections in place, researchers can be widely dispersed geographically. Such high connectivity is made possible by multi-gigabit-per-second data links. Researchers from industry, government, and academia are connected to one another and to other laboratories, both within the US and abroad; a global co-laboratory is achieved.

Software development begins to resemble a process of construction, with libraries and pieces of validated code being assembled to solve "grand challenges." The most exciting combination is the efficient melding of numeric and symbolic computing so that researchers can perceive the problem using both modalities of the brain: the analytical/linear as well as the visual/spatial. Hybrid architectures allow unique processing capabilities, perhaps geographically remote, to be joined to the researcher's own computational system, greatly expanding the types of problems that can be treated. Software development tools facilitate rapid prototyping of these hybrid architectures.

Rapid software prototyping is available for entire software codes and lower-level algorithmic approaches to problems. "Tools" are provided to determine architectural tradeoffs among computational methods as well as code validation. Artificial intelligence enables "intelligent assistants" to enhance the researcher's productivity.

The R&D environment becomes more connected with the business functions of the organization. R&D becomes better integrated with manufacturing, enhancing firm's flexibility in responding to markets. Particular manufacturing processes are also improved, minimizing waste and dramatically shortening production time. Marketing uses supercomputer-generated graphics as well as "movies" to communicate with potential customers. For example, reducing a competitor's products to its molecular definition. The detailed, supercomputer-generated visual form can reveal differences in chemical formulations, allowing marketing to point out "impurities" or manufacturing deficiencies.

The interconnection of researchers has increased the importance of computer security and network management, resulting in new legal challenges involving intellectual property and perhaps unfair competition. Databases are used routinely for configuration management, not only for crucial processes and parts (as is done today in the manufacture of complex products) but also for crucial software "modules."

## The Conundrum

Supercomputing is the leading edge of the computer industry.[12] Although supercomputing accounts for perhaps less than 10 percent of the industry's revenues, supercomputing pushes the technology for the broader computer industry as well as for important supplier industries. Supercomputing began as a uniquely American industry. The Japanese announced commercial supercomputer products in the early 1980s and a parallel computing thrust in the early 1990s. They are integrating the computational approach to science and engineering in competitively relevant industries more rapidly than in the US.

In the US, the supercomputing sector has been distinct from the overall computer industry. In Japan, the entire computer industry is vertically integrated. This difference in industry structure has already created intense competition. Such rivalry can be expected to increase.

In 1983, some 60 of what were then defined as "supercomputers" had been installed worldwide, mostly in the US. Approximately half of those were at government sites. By 1990 approximately 400 had been installed.[13] A key influence on the industry's growth will be increasing the application of supercomputing to industrial problems.

The number of supercomputers in industry has increased markedly, with this trend more pronounced in Japan than in the US (table 1). Supercomputers in Japan are concentrated in the electronics, pharmaceutical/chemical/biotechnology, and construction industries.

The Japanese companies that compete in supercomputers also supply the rest of the computer industry. These competitors are vertically integrated companies, leaders in the highly competitive semiconductor and disk industries as well as in computing. Japanese manufacturers are important sources of leading-edge chips and disks for the overall US computer industry—which includes the US supercomputer industry. In the US, however, such suppliers are treated as separate industries, barely related to computer manufacturing. America's fragmented perspective is vulnerable to the policy of any country that nurtures its information-technology sector as a whole.

Other nations have made catching up with and surpassing the US in computer-based technologies a priority, and have developed formal "information policies." Such competition often

**Table 1**
Adoption of computational approach to R&D.

| | United States | Japan |
|---|---|---|
| **Commercial applications** | | |
| Aerospace | 16 | 0 |
| Automotive | 6 | 12 |
| Chem/Phar/Biotech | 4 | 8 |
| Construction | 0 | 11 |
| Electronics | 2 | 14 |
| Energy | 2 | 4 |
| Petroleum | 19 | 0 |
| Service Bureau | 3 | 10 |
| Finance | 0 | 2 |
| Other | 0 | 1 |
| | 52 | 62 |
| **Government applications** | 90 | 12 |
| **University applications** | 31 | 30 |
| | 173 | 104 |

*Source:* estimates by Cray Research, Inc., January 1990. (Does not include IBM installations.)

ignores free-market practices.[14] In fact, the US is one of the few countries not to have defined an "information-technology policy" for the economic future, although there are many federal initiatives that affect the US computer industry.

Japan and the US are well matched on R&D expenditures as a percentage of GNP. However, with defense-sponsored R&D removed the US has actually been losing ground to Japan since the mid-1970s. Furthermore, the rate of civilian R&D investment as a percent of GNP has been rising faster in Japan than in the US since the mid-1980s.[15] A large proportion of US R&D for the overall computer industry is funded by the Department of Defense—by one estimate, as much as 80 percent in some years (see note 14). Usually such R&D only satisfies mission requirements, and has minimal overlap with commerce and markets. This is also true of current R&D efforts in high-performance computing. Too often, R&D is focused on revolutionary approaches to achieving performance, rather than on filling current technology gaps or extending current processes. In fact, current policy for advanced R&D under the auspices of the Department of Defense apparently precludes any support for simply extending today's state of the art. For example, the funding directive for R&D in high-performance computing architectures specifically states: "Program research should investigate innovative approaches and techniques that lead to or enable revolutionary advances in the state of the art. Specifically excluded is research which primarily results in evolutionary improvement to the existing state of practice or focuses on a specific system or hardware."[16]

By policy, US R&D investments (outside of mission-specific R&D) go primarily to national laboratories and universities—open environments intended to benefit everyone. US information policy has different approaches for government, university, and business. The private sector is primarily dealt with indirectly, through hoped-for spinoffs of R&D performed for government agencies or for the Department of Defense. With the increasing specialization of mission R&D, the future government research benefits for industry are less encouraging. Increasingly, R&D in Japan is performed in private laboratories. In studies of the relative abilities of American and Japanese firms to source technology externally and to translate that technology into commercial products,[17] Japanese companies showed a time advantage of 25 percent and a cost advantage of 50 percent compared to their US counterparts.

Sourcing external technology and developing technology internally are costly investments, not to mention the expense of ultimate commercialization. "There is indeed a problem in financing . . . the big D . . . [,which] involves taking the fruits of R&D to full-scale production and marketing. Innovation is not complete at the prototype stage. Moreover, the marketing element of the big D and its huge costs are as important as the science and technology."[18] In fact, full-scale development can be 100 times more costly than initiating the research. "Herein lies the problem, particularly if it takes a long time, which in the field of electronics, is usually the case. It takes heavy front-end money, with a long and risky payback—patient money."[19]

American R&D policy should continue to ensure commercially relevant high-risk R&D and should also encourage US companies to transfer, adapt, and commercialize externally sourced technology. US policy must also support an infrastructure that will enable US companies to compete. Traditionally, infrastructure has meant universities, capital markets, bridges, roads—the underpinnings of society. Increasingly, the term emphasizes a highly productive technology and manufacturing base—vital contributors to economic growth and leadership.

### Policy and High-Performance Computing

Impediments to wide use of supercomputers have included access, training, and cost. In 1984, the National Science Foundation began its Supercomputer Centers Program. At the beginning of the 1990s there were four national Supercomputer Centers at universities, as well as many state-supported Supercomputer Centers, offering access to state-of-the-art supercomputers and a well-appointed support system with a full complement of tools and support staff necessary for productive research. A recent survey by the National Association of State Universities and Land-Grant Colleges[20] points to the leadership role these centers have had in accelerating the computational approach to research. Federal policy continues to focus on ways to promote more widespread access to supercomputers. The current emphasis is on a broader inclusion of industry and on providing access to more experimental high-performance computing architectures.

In the early 1990s, both the Executive Branch and the Congress have articulated policy initiatives directed at high-performance computing. The importance of high-performance computing is emphasized by this unusual situation of having two branches of government agree on the focus of R&D. These initiatives are concerned with the long-term competitiveness of both the high-performance computing industry (which includes supercomputers) and the other technology-intensive industries. Semiconductors and software R&D, for example, are explicitly identified as crucial.

Current R&D efforts in high-performance computing are focusing on massively parallel systems and on other technologies (such as software) necessary to make these innovative architectures available for routine problem solving. The degree of parallelism that can be productively applied to a single problem is still a research question.

The goals of the US government's policy in the early 1990s are these (see note 8):

Maintain and extend the United States' leadership in high-performance computing, and encourage domestic production.

Encourage innovation in high-performance computing technologies by increasing their diffusion and assimilation into the American science and engineering communities.

Support US economic competitiveness through greater use of networked high-performance computing in analysis, design, and manufacturing.

Reinvest in US education to ensure that America will have a well-stocked pipeline of computational scientists to address the "grand challenges."

The US government proposes using large, currently intractable problems to focus the R&D in computer and computational science. R&D in four interrelated areas will be crucial:

high-performance computing systems, especially highly parallel architectures

advanced software technology, including algorithms, software tools, languages, compilers, utilities, and libraries focused on the needs of parallel architectures

deployment of the National Research and Education Network

improving the pipeline of scientists and engineers, and improving the educational system in general.

Computational science continues to build on US strengths in theory and experimentation, but equally includes computational approaches that can be expected to further stimulate science-based innovation. Achieving the broad benefits of the federal initiatives in high-performance computing will depend on the manner of implementation and on success in involving many researchers from a cross-section of technology-intensive industries.

Among the themes for supercomputing in the 1990s are multidisciplinary research and closer cooperation among researchers government, industry, and academia. Competitiveness will be fueled by rapid advances in computational science as well as in the science of parallel computation—and by the application of these advances into high-value-added products and processes. Computational science as an equal partner with the United States' traditional strengths of science and theory will accelerate the rate and the range of innovation.

The better use of time in design and manufacturing is increasingly important for US competitiveness. Computational science as an equal partner with the traditional US strengths of science and theory will accelerate and expand innovation. Applying supercomputing technology to industrial problems will increase the US standard of living, the ultimate measure of US competitiveness.

### Notes

1. The terms *very-high-performance computing* and *supercomputing* are used somewhat interchangeably. Generally, *VHP computing* is broader, including advanced research and development and experimental architectures as well as the current state-of-the-art supercomputers.

2. National Material Advisory Board Commission of Engineering and Technical Systems, *The Impact of Supercomputing Capabilities on U.S. Materials Science* (National Research Council, National Academy Press, 1988).

3. George Stalk, Jr., "Time—The Next Source of Competitive Advantage," *Harvard Business Review,* July-August 1988, page 41.

4. Steven J. Klin and Nathan Rosenberg, "An Overview of Innovation," in *The Positive Sum Strategy* (National Academy Press, 1986), page 7.

5. Wesley Iverson, "Detroit Plays up High Performance Engines in Race Against Japanese Supercomputer Horsepower," *Supercomputing Review,* November 1989.

6. See page 3 of the source listed in note 2.

7. See page 8 of the source listed in note 2.

8. *Federal High Performance Computing Program* (Office of Science and Technology Policy, September 9, 1989), page 1.

9. Jack Worlton, "Existing Conditions," in *Supercomputers: Directions in Technology and Applications* (National Academy Press, 1989).

10. Bruce H. McCormick, Thomas A. Defanti, and Maxine D. Brown, *Visualization in Scientific Computing* (National Science Foundation, July 1987).

11. Jack Worlton, "A Concise Technology Forecast for High Performance Computing in the Early 1990s," internal communication, Cray Research, July 1989. Gains in processor performance have, so far, been realized through technology improvements in silicon-based components (chips) and through moderate increases in the numbers of processors. For example, the Cray-1 logic chip was implemented in emitter-coupled logic (ECL) and had 16 gates per chip. The logic chip of Cray Research's top-of-the-line eight-processor YMP, announced in 1988, was implemented in ECL with 2500 gate arrays. Densities of 10,000–20,000 gate arrays are anticipated in future products, and alternative-technology implementations are under investigation.

12. Geoff Lewis, "Is The Computer Business Maturing?" *Business Week,* March 6, 1989.

13. "Update," *Computer Magazine* (IEEE), November 1989, pages 55–56.

14.  Benjamin G. Matley and Thomas A. McDonald, *National Computer Policies* (IEEE Computer Society, Computer Society Press, Ventura College, 1988).

15.  *Science and Engineering Indicators (NSF)—1987* (National Science Board), page 3.

16.  *Commerce Business Daily,* October 23, 1989.

17.  Edwin Mansfield, testimony before Joint Economic Committee of Congress, December 2, 1987.

18.  Robert Malpas, "Harnessing Technology for Growth," in *The Positive Sum Strategy* (National Academy Press), page 110.

19.  Ibid.

20.  Higher Education and Technology Committee of National Association of State Universities and Land-Grant Colleges, *Supercomputing for the 1990s: A Shared Responsibility* (Washington, D.C., January 1989).

# 3

## 2001: A Microprocessor Odyssey

*Patrick Gelsinger, Paolo Gargini, Gerhard Parker, and Albert Yu*

When the first microprocessor was built, in 1971, no one could have anticipated how widely it would infiltrate the modern world. The 4004 microprocessor was originally designed as a single-chip solution for a programmable calculator—one that would replace at least twelve separate chips. In time, it evolved into a device that was capable of controlling the operation of thousands of machines, appliances, and manufacturing processes, from the video recorder to the industrial robot. The microprocessor also became the avatar of the personal computer revolution, which in less than 20 years has transformed business, science, engineering, and everyday living. Today, it is found in workstations (advanced desktop computers with sophisticated math and graphics capabilities), file servers (systems that control networks of computers), and minicomputers.

In 1989 the microprocessor, for the first time, surpassed the million-transistor mark, with the introduction of the Intel i860 CPU (a central processing unit designed for high-power math and three-dimensional graphics applications). To make this chip, engineers placed more than a million transistors on a piece of silicon whose length was about the diameter of a dime. The consequences of this feat have only begun to manifest themselves. No one knows how many tasks software developers will program for it, or how many uses it will ultimately have, but its possibilities are vast.

The coming generations of processors will undoubtedly be even more astonishing than the last. In less than 20 years, the semiconductor industry has gone from the 4004 CPU, with its 2300 transistors, to the i860 and i486 CPUs. The latter, introduced in 1989, has 1,180,000 transistors—more than 500 times as many as the 4004. The i486 also *does* much more than the 4004. It performs several tasks at the same time (multitasking), and it can solve very complicated problems jointly with other processors (multiprocessing). (In our parlance, *microprocessor, processor,* and *CPU* are equivalent; *devices,* a word with many meanings and usages, will refer hereafter to the transistors and other electronic entities on the microprocessor.) We will speculate in this article about what features the microprocessor of the year 2000 will have and what it will do. We will find the first question easier to answer than the second.

### Where Technology Is Driving the Microprocessor

#### Line Widths

It is easier to predict how the microprocessor itself will change than how it will be used, because we can easily extrapolate such trends as the increasing number of transistors integrated on new chips since 1971. We also know how the widths of the lines etched in silicon have decreased, and how the size of the die of successive versions of the same processor (that is, the sliver of silicon that forms the chip) has shrunk with each decrease in line width.

At the same time that we can reduce a particular processor in size by decreasing line widths, we can build more advanced new processors by using more transistors. This causes an increase in the size of the die of the most advanced processors from year to year, another variable that we can measure. New developments in process technology (the methods used to engineer ever-denser silicon chips) and in the technology of manufacturing those chips in volume hint at how far these trends can continue.

Figures 1–3 convey some idea of how radically microprocessors have become miniaturized. One step in creating transistors is to etch lines in silicon wafers to form the outlines of circuits. This is done by sending electromagnetic radiation—such as visible and ultraviolet light—through a mask (a kind of circuit blueprint possessing the patterns of the desired circuits). The narrower the lines forming transistors, the greater the number of these devices that can fit onto a single die, and the faster these circuits will carry out their appointed tasks.

Figure 1 shows that in less than 20 years line widths have slimmed down from the diameter of a hair to slightly less than the length of the average bacteria, 1 $\mu$m. Process engineers have narrowed these lines to smaller than the eye can see by

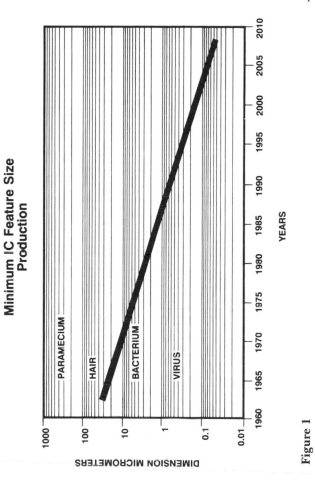

Figure 1

using electromagnetic radiation with shorter and shorter wavelengths, from visible to ultraviolet light and beyond. By the year 2000, the line width should reach 0.2 μm. A line width of 0.1 μm has been demonstrated in the laboratory. No one knows how much further this process can go. To manufacture transistors with 0.1-μm line widths, we will ultimately have to turn to x rays and electron beams.

### Denser and More Functional Chips
The narrower dimensions of transistors mean that we can cram more of them onto a given area. By extrapolating from data about previous microprocessors, we can guess that the area occupied by a single transistor in the year 2000 will be a thousand times less than what it was in 1971. At the same time, the industry has been adding more and more transistors to larger and larger dies. From our chart of the increase in the size of microprocessors through the years (figure 2), we can predict that the processor we call "Micro 2000" will be about 1 inch square.

While process engineers are shrinking transistors, manufacturing engineers are decreasing the number of defects per square centimeter of die. The semiconductor industry is slowly decreasing the number of defective dies in a batch (currently hovering around 200 defects per million), and can be expected to achieve a defect rate of nearly zero.

With smaller and smaller transistors and larger and larger dies, how many transistors could we potentially put on Micro 2000? The answer lies in figure 3. The lower line represents the number of transistors on the dies of some prominent microprocessors. Since the number of transistors on memory chips is traditionally greater than the number on microprocessors at any given time, and since microprocessors incorporate more and more on-chip memory, we have plotted this line as well. In both cases, the number of transistors on a chip in a given year is about 1.4 times the number from the previous year. If this trend continues, as we anticipate it will, Micro 2000 will have more than 50 million transistors. And if microprocessors continue to integrate more and more memory, as we anticipate they will, the microprocessor line may well edge closer to the memory line of figure 3. Thus, the count could reach up to 100 million transistors on a single chip.

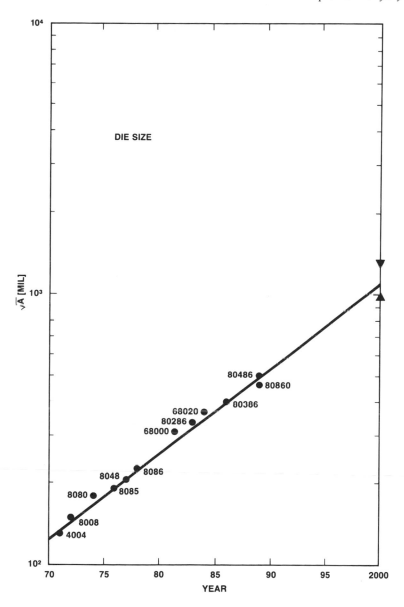

**Figure 2**

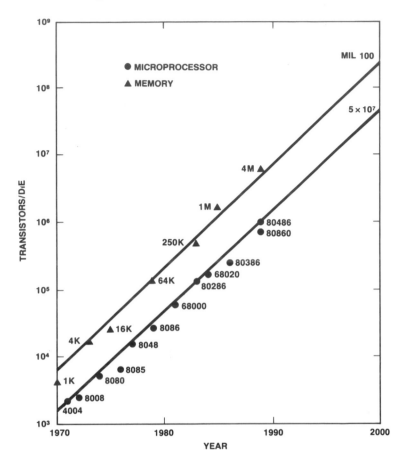

**Figure 3**

Driving down the size of Micro 2000's features has another benefit: electrical current can be driven through the chip at higher frequencies or "clock rates." Frequency is a rough measure of how rapidly information can be processed at a given time. Advanced microprocessors today commonly run at frequencies of 25–40 megahertz. We predict that Micro 2000 will have a clock rate of 250 MHz or more.

### The Features of Micro 2000

Another token of the age of computers and semiconductors, and one that is related to the pace of device miniaturization, is ever-growing functionality. With more and more transistors integrated onto chips, the functionality of those chips grows; in other words, they become capable of doing more for their users. The size of the chunk of data that a CPU can process at a time (for example, 32 bits) increases, and single chips integrate more and more of the functions of a complete computer system.

### Microprocessor Features: A Brief History

*Word length* refers to the size of a unit of data, in bits, that a CPU can process at a time. The larger the word length, the faster the data is processed. The first processor, the 4004, had a four-bit word length. As CPUs evolved, word lengths expanded to eight bits, then sixteen. Now 32-bit word lengths are commonplace, and Intel's i860 CPU has a 64-bit word length.

Another trend is the migration of features from mainframes and minicomputers to personal computers, and from other chips in the system to the system's microprocessor. For example, thoughout the 1970s mainframe computers had support for memory management, the task of assigning data to memory in the system so that programs can get at the information they need when they need it. This function first migrated onto a microprocessor in 1982, with Intel's 80286 processor. In the late 1970s, Digital Equipment Corporation's VAX minicomputer was the first to have a function called *paged virtual memory*, the breaking up of blocks of memory into smaller units (called *pages*) for more efficient use and storage. This is now a function of virtually every commercial processor.

Based on this trend, let us propose a new computing principle, the "law of migrating technology": Every good idea proved on the mainframe or the minicomputer will migrate to the microprocessor.

The larger transistor budgets of more advanced processors have let designers bring other functions on-board the microprocessor in order to handle particular kinds of problems and enhance overall speed. An example of this is the ability to do floating-point arithmetic. This function is essential to scientific and engineering applications, which require high numerical accuracy. In 1980 Intel introduced the 8087 math coprocessor, a separate chip designed to perform floating-point arithmetic when mounted near the microprocessor. Floating-point units have become resident on more recent microprocessors, such as Intel's i486 CPU and 960 embedded processor and Motorola's 88000 CPU.

Two other functions once separate from the processor are the data and instruction caches, memory units designed to store data and instructions used frequently by the processor. An on-chip instruction cache was first seen on the Motorola 68020, and an on-chip data cache on the 68030. These were initially quite small (256 bytes), but they are now quite substantial (8 kilobytes on the Intel i486 and i860 processors).

There are very few mainframe or minicomputer architectural features that cannot now be found on microprocessors. The most advanced processors of today are truly mainframe computers on a chip, in the sense that all of a mainframe's capabilities exist on the chip.

### Microprocessor Trends

Nothing in the semiconductor industry is more astonishing to the engineer and the general public than the continuous progression of device miniaturization. Still, to engineer a useful Micro 2000 will require the resolution of problems other than that of shrinking devices. Will Micro 2000's architecture run the vast software bases in use today? Since device miniaturization alone is insufficient to bring the performance gains we predict, what other innovations are necesary? What new capabilities will be integrated onto the chip?

### Compatibility Trends

Not every microprocessor can run every piece of software ever written. Throughout computing's history, processors have existed as families. From one generation to the next, some or all of a user's investment in software, other hardware, and training can be carried forward to a family's next generation. Moving to a new, incompatible processor requires redesigning old software to run on the new CPU (a process called *porting*) as well as redesigning the configuration of other hardware in the computer system. Depending on the scale of the change, this may incur two years' worth of high personnel and retraining costs or more. We believe that only a twofold or greater improvement in system performance makes a switch to incompatible hardware worthwhile. In most cases, the performance increases designed into successive members of the same family of processors are more than enough to compensate for any possible gains from switching. Thus, it appears that the general-purpose microprocessor families that dominate today will continue to do so through the next decade.

The only exception to this trend might be special-purpose architectures designed to solve certain kinds of problems, such as three-dimensional graphics and specialized math functions. To fill special needs, designers will extend the architectures of processor families, or design new processors especially suited to the need. An example is Intel's i860 processor, whose math and graphics units are designed to run complex computer models of scientific and engineering problems.

### Performance Trends

Microprocessor performance improvements will come from three areas: process technology, innovations in single-processor architecture, and multiple processor architectures.

Improvements in process technology—device miniaturization—will increase the speed of processors by about 20 percent per year. Figure 4 shows how various microprocessors have shrunk over the years as process technology has succeeded in reducing line widths. For each year shown, the size of a processor introduced that year is depicted along with the most advanced version of the previous generation. The die sizes of newer versions of processors tend to shrink because engineers manufacture them with transistors having smaller line widths.

Clock speeds of 250 MHz or more will require further improvements in processor architecture, such as integrating multiple microprocessors on a single die to create a "multiprocessor microprocessor."

Figure 5 depicts the increasing power of microprocessors, single-processor minicomputers, and mainframes, as measured in millions of instructions per second (MIPS). Microcomputers containing a single processor have already surpassed some minicomputers by this measure of computing power. As a result, many minicomputers, such as the Sequent Symmetry and the DEC MV3600, use microprocessors as processing elements. In the mid-1990s, microprocessors should also exceed single-processor mainframes in power. By 2000, a single processor should be able to run at 700 MIPS.

To boost this performance even further, designers will turn to multiple processors that cooperate. Since the late 1970s, mainframe computers have used multiple processors to boost

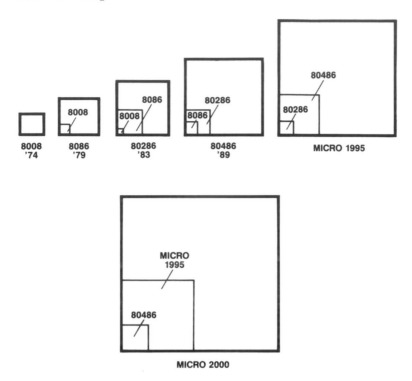

**Figure 4**

system performance; minicomputers have followed the same tack since the early 1980s. In light of our proposed law of migrating technology, we postulate that future processors will consist of multiple processors on a single die. In other words, if today's CPUs are mainframes on chips, tomorrow's CPUs will be multiprocessor mainframes on chips.

Research on the architecture of single processors is the third area that will yield performance improvements. Innovations here should lead to gains in speed as high as 50 percent per year. For example, the number of clocks (the basic unit of chip-level time) that pass during the execution of an instruction has been dropping steadily with successive generations. The one-clock-per-instruction barrier has already been broken with the Intel 1960 CA; it could eventually fall to half a clock per instruction or less.

We expect processors with 32-bit word lengths to dominate the next decade. Will Micro 2000 move beyond this? Intel's i860 processor already has—most data paths in this CPU are 64 bits

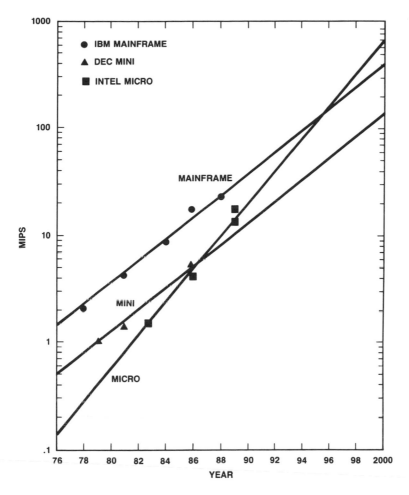

**Figure 5**

wide. Because the length of programs and the density of memory on CPUs continue to grow, we believe that Micro 2000 will be a complete 64-bit processor. This word length is ideal for solving many classes of problems with high accuracy. Internally, the growing length of new applications programs might require even longer word lengths. In order to transfer data from on-chip memory to the math units of the microprocessor rapidly enough, it would not surprise us if certain data paths within the chip were designed with 128-bit or 256-bit word lengths.

### Integration Trends: The Gateway to New Applications

In our attempts to imagine the microprocessor of the year 2000, we have gradually been moving from firm ground—a large base of knowledge and some fairly straightforward extrapolations of past trends—to softer ground. Extending the line on a graph of increasing transistor density is easy. Imagining the new techniques that will drive processor frequency to ten or more times the present speed is more difficult. Predicting what software engineers will program a 250-MHz processor with 50 to 100 million transistors to *do* is the hardest of all.

Flatly trying to predict Micro 2000's future is a problem best tackled by futurists and writers of speculative fiction. Instead, let us examine what types of functions might be integrated on this CPU, and what applications those functions might make possible.

Micro 2000's designers could build a processor suited to any applications they wanted—applications as generic as those of general-purpose computers or as specific as those of a machine dedicated to running computer models of natural phenomena. We will consider two directions: building the highest-performance microprocessor possible and building the most highly integrated personal computer on a single chip. The processors we propose have about 50 million transistors each, but individual units on each processor could be scaled up to create 100-million-transistor behemoths.

***The High-Performance Unit***    Figure 6 is the plan of one possible high-performance processor. This design is ideal for supercomputers that generate complex mathematical models. It is a multiprocessing mainframe on a chip, with four processor units, labeled CPU1 through CPU4. Each of these CPUs has

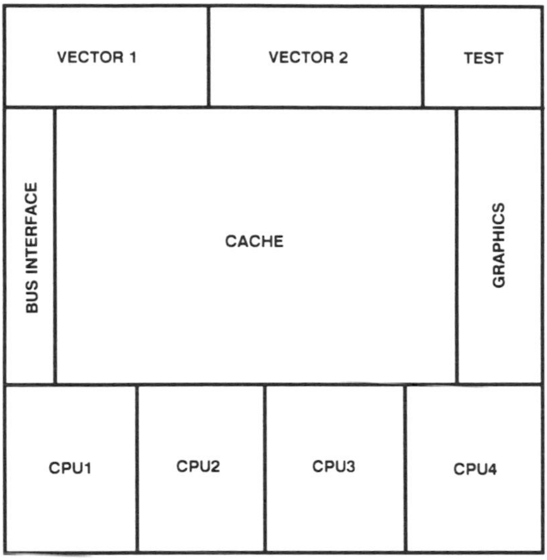

| VECTOR 1 | VECTOR 2 | TEST |
|---|---|---|

(Figure diagram: BUS INTERFACE, CACHE, GRAPHICS, CPU1, CPU2, CPU3, CPU4)

| Function | Number Units | Transistors Each Unit | Total Transistors |
|---|---|---|---|
| CPU | 4 | 4M | 16M |
| Vector | 2 | 4M | 8M |
| Graphics | 1 | 4M | 4M |
| Bus Interface | 1 | 2M | 2M |
| Test | 1 | 3M | 3M |
| Cache | 1 | 20M | 20M |
| **Total** | | | **53M** |

**Figure 6**

about 4 million transistors. (Recall that the largest CPU that exists today is Intel's i486, with about 1.2 million transistors.) Each CPU incorporates the improvements in single-processor architecture we touched on above; each includes a unit specialized for the floating-point mathematics required by many technical problems. The word length of each CPU is 64 bits. Individually, each can perform 700 MIPS; the four-CPU parallel system can run at 2000 MIPS.

There are also two "vector units," with 4 million transistors each, on this chip. Vector processing is a technique used to speed up numerical calculations by operating simultaneously on groups of numbers rather than on single numbers. Supercomputers use this method to perform hundreds of millions of floating-point calculations per second, and the i860 microprocessor supports this function, too. The two-unit system should be able to perform a billion floating-point operations per second.

We have also included a special-purpose graphics unit dedicated to displaying and printing high-resolution color graphics in three dimensions. Scientific and technical problems are easier to solve when they can be visualized. The same goes for business applications: programs that analyze large financial databases, such as options trades and acquisitions, are easier to understand when their output is displayed visually.

Other units on the chip support the processing units. The cache unit, physically the largest, includes 2 megabytes of memory (compared to the 8 kilobytes on the i486 CPU) for storing frequently used instructions and data.

***The High-Integration PC***    Our other design direction leads us to the complete personal computer on a single chip. Today, a PC based on the 275,000-transistor 386 CPU might require additional logic chips throughout the rest of the system, amounting to 250,000 transistors. These chips control such functions as floppy-disk and hard-disk drives, the system's memory, and its ability to communicate with other computers.

The Micro 2000 in figure 7 incorporates all these functions on the portion of the chip labeled "miscellaneous PC." We have assumed that the number of transistors on logic chips other than the CPU in the PC of the year 2000 has grown to eight million transistors. This is 16 times what it is today. As in the high-performance version, this Micro 2000 has a large cache unit, and a graphics unit. It also has multiple CPUs and a vector unit—both half as many as in the last example.

Prominent on this version of Micro 2000 is an area labeled "human interface," for which we have set aside 8 million transistors but which could very well require many more. This unit is designed to make the PC "smarter," "friendlier," and easier to use than ever. At present, it is a black box. Exactly what functions it will contain are enshrouded in thicker fog than any we have encountered in this speculative exercise. We have come face to face with the two biggest questions that both designers and users of computers have wrestled with since these machines were first created: What will we use them for? How will they affect our lives?

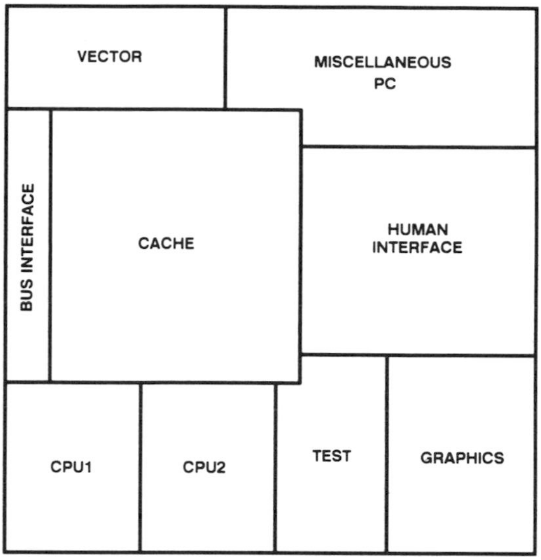

| Function | Number Units | Transistors Each Unit | Total Transistors |
|----------|--------------|-----------------------|-------------------|
| Processor |  |  |  |
|    CPU | 2 | 4M | 8M |
|    Vector | 1 | 4M | 4M |
|    Cache | 1 | 12M | 12M |
| Graphics | 1 | 4M | 4M |
| Bus Interface | 1 | 2M | 2M |
| Human Interface | 1 | 8M | 8M |
| Test | 1 | 3M | 3M |
| Miscellaneous PC | 1 | 8M | 8M |
| Total |  |  | 49M |

**Figure 7**

## Applications: Beyond 2000

Around 1980 Gordon Moore, chairman of Intel Corp., took a walk through his house. He was looking for potential domestic applications of microcontrollers. These modified microprocessors are specifically designed to control the operations of different kinds of machines, operations that were often being controlled by mechanical means. He found 84. In 1987 Moore returned to his list of applications to see how many of them had actually become microcontroller-based. Since Moore's first pass through his residence, microcontrollers had appeared in hundreds of different applications, domestic and otherwise. In the household realm, however, he found that fewer than half of the applications he could now count were ones he had originally predicted. He tells this story to illustrate the difficulty of predicting the future uses of a semiconductor product.

This leads us to propose that there are two classes of applications to which Micro 2000 will be adapted: the ones we can foresee now and the ones we can't.

A more useful division might be between doing what microprocessors and microcontrollers do now—only better—and doing what can be imagined now but cannot yet be done. To the former category belong a multitude of possibilities, including faster searching through stored information, more efficient control of manufacturing, and bigger, faster, easier-to-use networks of computers. In the scientific and technical realms, a high priority is placed on developing better computer models of extremely complex phenomena, such as the aerodynamics of advanced aircraft, the earth's climate, the economy, genetically engineered drugs, and the structure of the universe. Advanced computer models seem to eat up all the computing power that hardware engineers can throw at them.

More exciting to the imagination, but harder to discuss, are the applications that fall into the second category—including the "smarter, friendlier" user interface, which requires the successful implementation of several types of artificial-intelligence functions and full digital television. Rather than describing an exhaustive list of Micro 2000's applications, let us consider a few of the more interesting ones.

### Doing It Better

The second version of Micro 2000, the "PC on a chip," could mean a computer with the power of a multiprocessing mainframe in a box the size of a hand-held calculator. The first microprocessor, the 4004, was designed to put all of a calculator's logic on one chip; Micro 2000 would do the same for all the logic of a computer.

What could we do with this box? One option is to store prodigal amounts of information and search it instantly. Business executives and students need this, as does anyone without much time but with the need for the right answer right away. If we assume that one page of a book fills 2 kilobytes of memory, and that a typical book has 200 pages, the pocket computer could reasonably store the equivalent of 500 books. Operating at 700 or more MIPS, the system's quick-search algorithm (a feature of its advanced user interface) could find any piece of information in those volumes in half a second.

If we wanted to have more information available to us—say, what is stored in the 8000-volume neighborhood library—a 16-chip system, slightly larger but still portable, would do. If the average book is about an inch thick, we could carry around more than 650 linear feet worth of books—and knowledge whose value can't be measured.

### Doing What We Can Only Imagine

Today a user must learn to speak a computer's language in order to use it. To make it easier to use, software designers seek to make the computer speak the user's language. Although they have already made some progress toward this goal, there is still some distance to travel before the computer interfaces with the human the way humans interface with one another: by spoken language and visual cues.

Micro 2000 will be only a part of the solution, the hardware part. Developing a system capable of natural-language processing and computational vision is the extremely challenging software-based part. The natural-language system function must not only be able to recognize words that do not sound the same when spoken by different people; it must also be able to "interpret" what has been said and to respond in kind. The computational vision system must not only distinguish among distinct objects in its visual field; it must also analyze what the scene "means," and respond appropriately.

Long before the field of artificial intelligence achieves its breakthroughs, other technologies promise that a "smarter, friendlier" interface will not be long in coming. Full-motion, real-time, interactive video is just such a technology. A version of this under development at Intel is called Digital Video Interactive (DVI) technology. This system allows the PC user to capture moving and still video images, to store and manipulate them on optical disks, and to create video programs with full-motion and still images, text, and stereo sound. DVI technology's applications include simulator training and education. A teacher might run a program that puts students in a foreign country's capital city, to teach its culture and language with images of its people and historical sites. Sitting at the system's screen, students would run the program themselves. They would decide where in that city they wanted to go, and what they wanted to see and hear, by directing their motions through the video program with a joystick. A narrator would explain what they saw. People would pass, greet them, and stop to talk in the native language. The students could visit any or all of the "places" in the DVI program's database, selectively accessing information rather than being limited to a static, unvarying videotape.

Intel is integrating DVI technology onto a set of custom chips that will be ready around 1992. Micro 2000 will be large enough to permit designers to integrate full DVI-technology capability as a single unit on the chip. DVI technology requires more computing power to convert an image into a form storable in the system than it does to play the image back—power that Micro 2000 could furnish.

Another application for Micro 2000 is full digital television. This contrasts with high-definition television (HDTV), the improved broadcasting technology that several nations' electronics industries have begun developing. HDTV improves picture quality without altering the analog signal broadcast from the TV station. Digital television goes one step farther, broadcasting television signals digitally. In an all-digital format, there would be no loss of signal quality from radio interference, electrical storms, or sunspots, all of which prey on present-day signals. Micro 2000 is ideally suited to this technology, which, like DVI technology, requires enormous amounts of processing power. Still farther down the line is three-dimensional, holographic television, a technology that could draw on Micro 2000's 3-D graphics power.

### The Artificial Companion

The social consequences of Micro 2000's computing power—in particular, the changes that a "smart, friendly" interface might cause in the way we live and think—cannot be adequately treated in a short chapter. We think it is worth noting, however, that information is regularly misinterpreted in this age of mass media and information abundance. More powerful processors with powerful graphics will make it easy to display information visually rather than numerically, and therefore easier to interpret the information. Chip-scale PCs with smart user interfaces will enable their users to become active seekers of information, rather than the passive absorbers we are encouraged to be by nonselective media such as television. With such a machine in our pocket or our briefcase, we might become more active decision-makers, more the masters of our respective fates.

And suppose that machine were intelligent. A thinking companion could augment our mental capabilities in unimagined ways. Around 1976 the semiconductor industry produced its first "brain." The sum total of all the transistors ever manufactured on semiconductors that year roughly equaled the number of neurons in the human brain—about 100 billion.

With 100 million transistors in Micro 2000, it would take only one thousand processors to produce a computer with one "human brain equivalent" of transistors. Intel has already embarked on a project, called Touchstone and partially funded by the Defense Advanced Research Projects Agency, to build a prototype of a massively parallel computer based on the million-transistor i860 processor. Such systems could ultimately use up to 2000 processors. Replaced with the 100-million-transistor Micro 2000, a single Touchstone system would have as many transistors as the brain has neurons. With the dream of an artificial companion that much closer to reality, it may soon come time to ask: When will we put the first brain on a single chip?

# Wealth and Mastery

# 4

## Knowledge and Equality: Harnessing the Tides of Information Abundance

*Harry Tennant and George H. Heilmeier*

Perhaps it gets tiresome to read, as we have read for years, that advances in computing are going to change the world. But it's true. They already have, and it will continue. The changes often come as revolutions rather than as gradual adjustments. The most recent revolution occurred during the 1980s and involved pervasive personal computers and radically easier-to-learn application software. As a result, personal computers are now commonplace in the modern office—familiar as pencils and paper, telephones, paper clips, and calculators. They are used as information power tools.

There is another revolution coming that will change the way we think about computers and the way we think about our work. We will come to think of computers as important sources of answers and insights, fulfilling part of the role played today by colleagues, advisors, file cabinets, blackboards, libraries, and experience. Rather than simply giving us easier ways to conduct analyses, computers will be helping us to determine which analyses need to be done. Rather than simply providing tools to calculate a budget, computers will be pointing out the places where the budget needs more attention and where it is inconsistent or out of line with respect to industry forecasts. The eventual benefit of information abundance is not simply that there will be more information available in a library or in electronic form but that the information and knowledge that is available will be applied to problem-solving as readily as adding a column of numbers or moving a paragraph of text is done today.

Twenty years ago, a programmer's first commandment was "Don't waste machine time!"—it was frightfully expensive; we were billed by the fraction of a second. We began the 1980s heralding the coming age of "computational plenty." It is here.

Today, after dramatic reductions in cost per function, the imperative is to find more for computers to do for us. There is still plenty of room for more computer sales, and computing power will continue to increase while costs continue to fall; however, there are over 12 million personal computers, and they sit idle on the average of 23 hours per day. This is not to say that they are necessarily bad investments, only that there are plenty of computing cycles around for those who want them.

Computational plenty has arrived on schedule. And computation will get more plentiful. Processing speed has increased by a factor of 100 since the mid 1960s, while the costs of processors have fallen by a factor of 10,000. During the 1980s, these trends led to pervasive desktop computing and to the embedding of computers in factory machinery. Although we are approaching some fundamental limits in semiconductor technology, it is likely that the next 25 years will bring similar advances.

*Expect a radical change in microelectronics in the late 1990s.* The increasing power and the decreasing cost of integrated circuits—the silicon chips that are the essential components of modern electronic systems—have fueled an unprecedented rate of technological change since 1960 and put us on the verge of an age of information abundance. This change has occurred because it has been possible to reduce the minimum size of individual circuit elements, to increase the area of chips, and to increase the efficiency with which electronic devices are arranged on a chip. Since 1959, minimum element size has decreased 11 percent per year, and chip area has increased 19 percent per year, while the number of devices that could be placed on a chip of a given area has doubled every 10 years. Together, these three factors have produced a 100-fold increase per decade in the number of electronic devices that could be put on a single silicon chip—an achievement that has been the most important factor in increasing the functionality and decreasing the cost of integrated circuits.

Integrated-circuit technology is still advancing exponentially. The density of components and the performance of circuits are still improving rapidly, and cost is falling at the same rapid rate. The good news is that such density, cost, and performance improvements are going to continue. The bad news is that this trend may not continue at the same rate into the late 1990s if there is nothing more than an extrapolation of current technology.

Most experts agree that the exponential decrease in the minimum linewidth of optical lithography (the current method for "printing" circuits on silicon) will probably reach a limit of about 0.25 μm in the 1990s. (For comparison, today's 1-megabit dRAMs have a minimum linewidth of about 1 μm.) Although other circuit-"printing" technologies (e-beam, ion-beam, and x-ray lithography) can decrease the minimum linewidth below such limits, at those dimensions other factors limiting the scaling of ICs come into play (such as interconnect capacitance, channel capacitance, interconnect resistance, parasitic resistances, ionizing radiation, punch-through, hot electron phenomena, and electrical breakdown). It appears that minimum linewidths for high-volume ICs will not get much smaller than 0.25 μm.

Larger chips and 3-D ("high-rise" or "trench") circuits are two evolutionary approaches to sustaining the exponential trends of microelectronics. However, larger chips imply larger, more expensive packages and more complex, more expensive manufacturing processes. In addition, patterning large chip areas with submicron features has limits in its own right. The use of 3-D circuits is a viable approach, but this merely prolongs the inevitable for a few more years, and it also increases manufacturing complexity and costs.

The long-term solution may be to find an alternative to transistor-based microelectronics. An alternative being investigated at Texas Instruments under the leadership of Robert Bate is called "quantum-coupled devices." Whereas transistors behave as simple switches, require extensive interconnection wiring, and are close to their fundamental size limitations, quantum-coupled devices lend themselves to more complex functions, simultaneously increasing the density of functionality on the chip and reducing the need for interconnection wiring. Furthermore, quantum-coupled devices can be made much smaller than MOS or bipolar transistors—in fact, the largest quantum-coupled devices may be about 100 times smaller than today's typical transistor.

In addition to advances in processing, advances in information availability are likely to carry us across a threshold so that we begin to view computers differently. By the middle of the 1990s, the typical storage capacity of a personal computer will have increased by a factor of about 100, to about a gigabyte. And local networking will have been made sufficiently transparent that the information your co-worker chooses to make accessible will be as available as if it were on your own computer. This will effectively multiply the available information by the number of people on the local network—about another factor of 100. This means that the information available through one's computer will be about 10,000 times what is typically available today—about the difference between having one book and having a community library. Add to that the emergence of nationwide computer networking, eventually making computers as accessible to one another as telephones are today. The step from national networking to global networking has already been taken for financial and military transactions. Next add the availability of information utilities providing access to current and collected periodical text in electronic form. As these technologies make information more available through our computers, we will think of computers less as processors and more as information sources. We will see them as more like libraries than like calculators. However, all that information will be subject to automatic analysis, decision making, simulation, and visualization. As someone has said, we will soon look back on the libraries of today and be amazed that the books didn't talk to one another.

We will soon routinely access the depths of vast oceans of collected data and the flow of thundering rivers of new information. And the material will be in all sorts of forms: databases, spreadsheets, text, graphics, maps, simulations, models, audio, video. The "information explosion" is a familiar theme, but what we now see is something very different, more extensive and more managable: information and data in an electronic form that will automate and enhance our way of dealing with it. And the way to make use of data and information is to apply knowledge to it.

*Expect ease of use to reemerge as a major issue in computing in the early to mid 1990s.* Ease of use was one of the hot topics in computing in the early 1980s. Along came the Xerox Star, the Apple Lisa, and (most important) the Apple Macintosh, and the problem was "solved." Ease of use was no longer a topic of intense interest. But it was shown to be an effective investment of computing resources. In the 1950s, by our estimate, less than 5 percent of all CPU cycles were devoted to the user interface; today it is probably close to 50 percent and continuing to increase. CPU cycles increased at annual rates of about 25 percent through the 1970s and about 45 percent in the 1980s, so the size of the CPU "pie" has been increasing along with the percentage of cycles devoted to interface tasks. However, one way the ease-of-use problem was "solved" was by stripping applications down to something that could be mastered easily and quickly by a new user. Instead of database systems, users were given table managers; instead of full constraint-propagation, they were given spreadsheets with numerical-only computing capability and one-way constraints. These applications fit the hardware as well as the conceptual readiness of the users. But a new situation is developing.

In the early 1980s many people wouldn't touch the strange and forbidding devices that were computers. Not only were they unfamiliar, but one wrong move could crash programs or destroy weeks of work. Since computers have gotten much easier to use, many more people have become comfortable with them. A wealth of new kinds of applications have come available. The power and the capacity of computers have continued to increase, doubling every couple of years. Computing power is ready to support much more sophisticated applications than it could in 1980 for the same cost, and the many people who have grown accustomed to computers are prime customers for more powerful software systems. For many users, the computer has become practically invisible; they see only the applications that run on it.

The early 1980s' solutions to problems of usability—which still dominate—are strained by the requirements of moving to relational database systems, large-scale information retrieval systems, simulation systems, animation systems, 3-D graphics systems, or systems that permit hundreds of operations on thousands of kinds of objects. The menu, the scroll bar, and the dragged icon, still useful today, will not be adequate to the kinds of applications that will appear by the mid 1990s. The success of these more complex applications depends upon how usable they can be made.

In the mid 1980s, ease of use came to mean direct manipulation (menus, What You See Is What You Get, draggable icons, point-and-click). Notice, however, that the biggest computer installations in the world, the IBM IMS-based corporate information systems, have gotten through the ease-of-use revolution of the 1980s virtually unchanged. As personal computer systems gain in local storage capacity, and as the availability of information is increased by networks and dial-up on-line data services, applications of personal computing will come to look a lot more like today's mainframe information systems. These personal systems will have access to vast amounts of information.

The emerging problem of usability is a more conceptual problem. It is a problem of extent—of having some knowledge of the existence of thousands of different things. It is like the problem of knowing what information is likely to be in an almanac, in the *Statistical Abstracts,* in *Fortune* magazine, on the National Portrait Gallery's video disk, or in a "National Geographic Special." It has more in common with library card catalogs, book indexes, *TV Guide,* and abstracting services than with selecting dinner from a menu. It has more to do with an intelligence agency integrating information from sensors, from the press, and from operatives around the world than with arranging papers on a desktop. The coming ease-of-use problem is one of developing *transparent complexity*—of revealing the limits and the extent of vast coverage to users, and showing how the many known techniques for putting it all together can be used most effectively—of complexity that reveals itself as powerful simplicity.

Not only is the technology developing to enable us to apply knowledge to abundant information, but there is a strong need for it. The percentage of knowledge workers in the American workforce—those who apply knowledge to create, modify, and distribute information (managers, administrators, engineers, lawyers, accountants, loan officers, secretaries, salespeople)—has been increasing since 1960, or even longer. Knowledge workers now constitute about 60 percent of the American work force. Because of their large number, their productivity has a strong effect on the overall productivity of the society. In other words, improvements in our standard of living will, in large part, depend upon knowledge workers' doing what they do more efficiently and effectively. But the productivity of knowledge workers has been declining.

The economist Lester Thurow reports that America's real business output grew by 18 percent from 1978 to 1985 while the blue-collar workforce declined by 6 percent to 30 million and the white-collar workforce rose 21 percent to 58 million; therefore, the respectable improvements in blue-collar productivity were dragged down by poor white-collar productivity. Steven Roach, of the investment firm Morgan Stanley, notes that blue-collar productivity rose by 13 percent from 1978 to 1984 while white-collar productivity fell by nearly 10 percent, yet more was invested in improving white-collar productivity per worker than in blue-collar productivity. (*The Economist*, December 13, 1986, p. 71)

Many consider these analyses of the importance and the poor productivity of knowledge workers controversial. Measuring the products of knowledge workers is much more difficult than measuring the output of factory workers has been. A similar argument is made with regard to counting the knowledge workers. With all that, an important point remains: a large portion of the workforce is primarily concerned with applying knowledge to create and manage information, and the computing and communications tools these people have been getting so far, though clearly useful, are not useful enough. But technology can contribute to a solution of this problem with new tools and automation for knowledge workers: *knowledge technologies.*

*By the mid 1990s, people can be expected to view personal computers as knowledge sources rather than as knowledge processors.* Today, we mainly think of computers as powerful processors: spreadsheet calculators, word processors, graphics renderers, simulators, controllers. Soon we will be thinking of them primarily as gateways to vast amounts of knowledge and information.

This prediction is based on two observations: First, knowledge work has become the dominant kind of work. The number of knowledge workers exceeded the number of production workers around 1980, and it continues to increase. Second, the infrastructure to support the storage, distribution, and processing of knowledge and information is developing. Personal archives will soon be commonplace, as will personal access to electronic libraries and electronically mediated news.

Today, we think of computers in terms of their speed. This is the essential quality of the process-centered view of computing. The essential quality of the knowledge-centered view of computing is *coverage*.

To understand the difference between the process-centered view and the knowledge-centered view, think of the difference between paper and pencil and a library. Both are tools to help us work with ideas. Paper and pencil are tools for recording and manipulating ideas; the ideas will be relatively few in number. We want the pencil and paper to work easily and reliably. We can even augment their capabilities for reorganizing and distributing ideas with scissors, tape, and a photocopier. Now think about why you go to a library. You expect to find ideas, and to use them to create or test new ideas. To be useful, the library must have resources in your area of interest. It must be well organized. It may not have the answer to your specific question, but you go there expecting to find ideas that can be modified or pieced together to get you farther toward a solution faster than if you had to generate all the ideas out of your own imagination. And more important, you will be working with ideas that have been tried and tested before; this tends to enhance the quality of the solution.

It is best to carry your pencil and paper along when going to the library. The library does not replace the need for paper and pencil. The tools to support processing are important there—they have become essential for analysis, for exploring possibilities, and for deriving new insights—but they must act on something; they are secondary to the information and knowledge. If you are commended on your solution and asked how you arrived at it, you may well say "I applied Green's technique to industry data for the last ten years." You would miss the point if you said "I used paper and pencil," even though you couldn't have applied Green's technique without pencil and paper.

### Knowledge Technologies

Artificial intelligence attacks a collection of problems: how do we get computers to reason, to see, to plan, to learn, to understand language? A variety of technologies have been developed to address these problems. The ones most widely applied today are what we call knowledge technologies—computer decision making and language understanding. Of these, far more applications of decision making, called *knowledge systems* or *expert systems,* are in use. With the amount of research being applied to learning systems today, it is quite possible that technologies for learning will be important for commercial application late in the 1990s. One final area, which has only loosely been associated with artificial intelligence but which has many characteristics in common with knowledge technologies, is *hypermedia.* The prospects for knowledge systems, natural-language systems, and hypermedia will be discussed below.

### Knowledge Systems Defined

Knowledge embedded in a computing system is the combination of structured information and an interpreter to apply that structured information to a problem. If documentation for a factory process is written on a text editor, it is machine readable, but it is not computer-based knowledge. If that documentation is then indexed to allow for better retrieval of relevant portions of the text, some structure has been added; it is more usable, but it is still not computer-based knowledge, because the computer is still not able to apply the knowledge. A person must do the interpretation. If, however, the information in the text is structured in such a way that a program (an interpreter) can use it to run the factory, then it is computer-based knowledge; it is executable knowledge.

Thousands of knowledge systems are in use today. Knowledge systems are being used in nearly every industry and nearly every function of business to improve consistency, productivity, quality, and speed. They are being applied to diagnostics, control, scheduling, decision making, and product configuration, and to making policies and procedures more accessible.

### The Importance of Explicit Knowledge

If we are to realize the advantages of knowledge technologies, the knowledge in knowledge-based systems must be available for scrutiny and review. This is one of the differences between, say, rule-based programming or object-oriented programming and conventional third-generation languages (FORTRAN, C, COBOL). Programming must be done in such a way that the knowledge is not lost in the details of implementation. In other words, having an operational knowledge-based program is only half the story; the other half is being able to readily review and modify the knowledge in the program.

### Benefits of Knowledge Systems

The developing technologies for the application of knowledge have four significant characteristics:

• They enable the user to apply knowledge without requiring that the user first understand it.

• They make possible the autonomous application of knowledge (decisions and judgment without human intervention), which permits the delegation of knowledge-based tasks to computers (as in disease diagnosis) and the application of knowledge where humans are not present (as in the case of intelligent spacecraft or weapons).

• They allow concise communication with humans and with other knowledge-based systems, relying on brief references to large knowledge structures that are shared by both conversants.

• They promise the autonomous creation of knowledge.

### Historical Perspective

These characteristics are of uncommon significance in the history of the uses of knowledge. To put them into perspective, consider the characteristics of three of the most significant knowledge technologies mankind has developed: language, writing, and printing.

***Language***    Much of language is used for emotional purposes or for a small repetoire of simple messages, in much the same way that chirping birds reassure one another of their presence and warn one another that they must keep to their own territory. Human language, however, has an additional component: the transfer of knowledge, which reduces the need for experience and discovery. If you go through an experience and thereby discover something new about the world, I can benefit from that if you just tell me about it. I do not have to experience and discover the same thing for myself. We are born with an enabling capability for language, but this is called a "knowledge technology" here because most of it—what we talk about, and how we express it—has been invented.

***Writing***    Writing freed knowledge from its constraints of space, time, and human memory. Because of writing, we can benefit from the thoughts of men who lived 3000 years ago.

We can also reconstruct thoughts of our own that we had three weeks ago and would have forgotten had we not written them down.

***Printing***   Printing made writing cheaper and thereby allowed it to become more pervasive. It is because of cheap printing that so much knowledge and information is available to us today—in libraries containing millions of titles, in daily newspapers, and even in junk mail.

It is clear that these knowledge technologies have had an enormous impact on the world. There have been the global-scale benefits, expressed in terms of the size of libraries and the quantity of knowledge being communicated in the world. There have also been personal-scale benefits: these knowledge technologies (and, likely, future ones) are crucial instruments of self-interrogation, self-discovery, self-organization, and self-realization.

The characteristics of computer-based knowledge enumerated above are of comparable significance to the characteristics of the other three knowledge technologies, and so one can expect computer-based knowledge to have a similar impact. It is worth noting, however, that Gutenberg's printing business was a financial failure. Computer-based knowledge technology will change the world, but it is not clear when this will occur or who will profit from it.

### An Analogy: The Effects of Cheap Paper and Printing

We can try to understand what the effects of computer-based knowledge technologies might be by drawing an analogy with the effects of inexpensive paper and movable-type printing, both of which became available in the 15th century in Europe. In Europe prior to the 15th century, *fact* didn't mean the same thing it means today. Before inexpensive paper and printing, what was called fact was what today we call opinion. Very little was recorded, leaving the chronicling of events and the results of what little measurement there was to the uncertainties of human memory. Life was conducted to a greater extent in the present, with less built on the past and with relatively little concern for the future. It is true that some people's memories were trained to be better than ours typically are today, but they were still subject to fading and modification. In addition, "facts" were often altered to fit the techniques of memorization: emotion was simple and exaggerated, and the telling was often in rhyme.

Commerce required the reliable recording of fact. It is a case of bootstrapping: commerce needed to record transactions before it could get very complex, and the availability of new technologies allowed the complexity to increase. The first major technology shift of this kind in medieval Europe was the adoption of techniques (originally Chinese) for the inexpensive manufacture of paper. The cost of paper fell by a factor of 4 in the 14th century. Cheaper paper meant more recording, which meant more copying, which meant a heavy demand on scribes, which eventually became a bottleneck for business. Then the advent of printing brought with it the widespread copying and distribution of fact. Printing technology quickly spread through the centers of European business.

The widespread availability of printed material fostered the emergence of better systems of organizing information and knowledge: the file, the catalog, the index, the bibliography. Also, the ready access to knowledge in printed form brought an end to unquestioned authority. Many attempts to reinstate unquestioned authority in society, whether political or religious, have also attempted to suppress printed knowledge through banning or destroying books. In other words, these technologies facilitated the externalization, the cross-breeding, and the wider application of knowledge.

### What Might We Expect from Knowledge Technologies?

Printing provided a means to transfer knowledge, but not without effort. First one had to learn how to read; then one had to read, understand, and internalize the knowledge found in books before one could apply it. And this put a practical limit on how detailed the knowledge would be. Today's knowledge technologies permit us to make use of knowledge without having first to internalize it, and the knowledge can be as detailed as is necessary since there is no problem of boring the reader. Today we find ourselves in the middle of a transformation of our understanding of what knowledge and information are all about. It will come to be more important to understand what knowledge is available and how it can be used than to become personally expert at the process of applying knowledge—just as today it is more important to know how to use the results of numerical calculations than to know specific algorithms for taking square roots, thanks to the pervasiveness of calculators.

### More Widespread Recording of Knowledge
The effects of knowledge technologies may be analogous to the effects of inexpensive paper and printing. First, the economic effects: the price of the medium (paper) declined by a factor of 4 in one century. We are seeing similar declines in memory prices, not in a century, but in about four years, and that has been repeated over many four-year cycles. The cost per copy fell by three orders of magnitude with the transition from scribes to printing. Falling costs of memory, processing, and communications will continue to make the production and the distribution of knowledge more economical. Knowledge specialists, such as accountants, physicians, and consultants, will greatly extend their reach, while purchasers of their expertise can expect the cost to fall. What we are seeing more of today, however, is the distribution of more mundane knowledge: how to repair a piece of equipment that rarely needs repair, how to schedule a factory, how to interpret policies and procedures. We are seeing many successful applications in areas where there is complexity but little incentive for a person to become an expert.

### A Change in the Nature of Decision Making
An intriguing possibility comes from the analogy between the change of meaning of "fact" and the current view of judgment. Fact was once subjective, not available for scrutiny, and rather mysterious, much as decision making and judgment are today. How was the decision made to introduce Product X? Why did I get a different raise than the person who sits in the next office? Just why is it that one investment seems more promising than another? Today we get vague answers to such questions, just as in the past one would get vague answers to questions like "When were you born?" "How many people live in this town?" "What is the average life span of Italian citizens?" Past attempts to make judgment formal and standard often resulted in inflexible bureaucracies. But a primary characteristic of knowledge technologies is the ability to deal with highly conditional situations, providing the flexibility that judgment in the world requires.

### Decision Making Open to Scrutiny

Perhaps most important, the decision-making process can be fully open to scrutiny, modification, and tuning. This is not to say that the structure of every decision will be made available to anyone who might happen to be interested; however, when it is appropriate for an individual to examine how a decision was made, the steps in the decision will be available for examination. There will still be disagreement over what knowledge to use, just as today there is disagreement over what data to collect to truly reflect a situation. (For example, there is a raging debate today over whether accounting data still reflect the health of a business, in light of the shift in the workforce from production to knowledge work.) But just as there is no debate today over what the data are (number of widgets sold, dollars paid for labor), knowledge technologies will make explicit what the knowledge is. And just as cheap paper and printing encouraged more widespread collection of data, contributing to the flowering of science in Europe over the following centuries, knowledge technologies will encourage more widespread articulation, collection, and distribution of knowledge. It is reasonable to expect that mankind's understanding of himself and of the world will improve dramatically over the next century.

As knowledge is articulated, we can expect it to become better organized and more readily utilized. We can expect it to be integrated into more of our activities, so that we will make use of the accumulated knowledge of society with less effort. These same organizational techniques should also assist us in developing new areas of knowledge, probably enabling us to develop knowledge and understanding in areas that heretofore have been overcomplex or have been represented by rare or obscure events in massive amounts of data. An example of this is a diagnostic system, built at Westinghouse, that consisted of a rule base that encoded the experience of problems with the assistance of an inductive component that attempted to infer rules from about 13,000 cases of malfunctions. The induced rules described how the defect rate depended upon a parameter that was previously thought to be entirely irrelevant to malfunctions: the temperature of a lubricant. This gave human experts reason to focus their attention on the new parameter. Subsequent study revealed that lubricant-temperature fluctuations did cause defects. The knowledge system had not only recorded knowledge and indicated a correlation; it had also assisted in the development of new knowledge and in the understanding of what that new knowledge was. Could this have been done without knowledge systems? Certainly it could have, but in the twenty years of operation of that production line the relationship had never before been noticed. And before this technology was available no one had been interested in manual examination of 13,000 cases of malfunctions. Had statistical techniques been applied to the same data, they would probably have indicated that a correlation of some kind existed, but they would not have been able to propose what the relationship was.

### *Experience from the First Decade of Commercial Application*

#### *Mundane Knowledge Dominates World-Class Knowledge*
Thousands of knowledge systems have been implemented and are in use today. Most of them are small systems running on personal computers, but some quite large systems have been built as well. A few years ago the general assumption was that the value of knowledge systems would lie in capturing the knowledge of "world-class" experts and making it available to the rest of us. Although this has been done to some extent, automating the application of more mundane knowledge has been pursued more vigorously. The experts are commonly local to a company, not "world-class," and the intent is not to be as smart as some brilliant person but to speed up everyday decision making, to be free from the drudgery of thinking, or to make the decision making more consistent over a large number of practitioners. Instead of generating insights to win Nobel Prizes, knowledge systems today are reducing work in progress through such applications as improving factory scheduling at Texas Instruments, automating the design of gas-processing equipment at BASF, making credit-authorization decisions at American Express and insurability judgments at Nippon Life, reading money-transfer telexes at a large New York bank, predicting and diagnosing malfunctions of power-generation equipment for Westinghouse, speeding up the repair of semiconductor-fabrication equipment at Texas Instruments, and reallocating US Navy ships.

### Answer Systems versus Insight Systems.

Nearly all the knowledge systems we see today were designed to give an answer to a question, such as "What disease does this patient have?" or "How can I fix this equipment?" Many uses of computing, on the other hand, are designed to lead the user to new insights rather than just hand him an answer; modeling, simulation, and visualization are examples. A numerical example is the difference between a calculator and a spreadsheet. A calculator can be used to add up monthly expenses, but a spreadsheet can be used to understand the implications of different patterns of spending and investing. They both apply the same knowledge (numerical calculation), but the spreadsheet puts it into a context that can be manipulated and "massaged" until the user comes to have a deeper understanding of trade-offs and possibilities.

People in many areas of science and engineering consider the computer as important, as a means for doing science, as the scientific method itself. "Analysis by synthesis," or coming to understand something by watching a simulation of it work, is used in fields from astronomy to airfoil design to the modeling of semiconductor devices. For the most part, we are not seeing this done with knowledge systems today. But we will. It is too useful not to appear, and non-numerical problems are too prevalent to be ignored.

### What Tools Are Needed?

Tools are on the market today which have been designed to eliminate some of the mechanical impediments between the user and the internal mechanics of knowledge systems. These tools are better than they were, but truly convenient knowledge programming has not yet appeared. We are still at a point in the programming of knowledge systems that is analogous to the point we were at when we could say that BASIC was easier to start with than FORTRAN. However, VisiCalc (the first commercial electronic spreadsheet) had not yet appeared.

The VisiCalc of knowledge systems will open up knowledge programming to large numbers of people. We all have little bits of knowledge that could be automated. At a lower limit, all those who use spreadsheets to understand numerical calculations could augment those spreadsheets with the knowledge to interpret what the numbers mean. For example, we see a mass of numbers in corporate balance sheets, and we know that those numbers could be used to describe many aspects of the health of the corporation they describe, but that step of inference is currently missing from most spreadsheets. When it is as convenient to enter as a numerical expression is today, such knowledge will tend to be collected and used, then examined and improved.

### Shared Knowledge and Shared Meaning

When we communicate, we are able to do so because we share a great deal of knowledge and information about the world. When you refer to "chair" or "profit" or "the second quarter of 1988," I understand you because my ideas of what those things are are pretty close to yours. That is the foundation of language, which is the foundation of culture.

Computer systems are in use today that extend that notion of shared knowledge and shared meaning in preliminary but suggestive ways. Insurance companies, for example, are using expert systems to make their agents more consistent—not individually better, but more consistent. The auditors of the accounting firm Coopers and Lybrand use a knowledge system to make their audits more thorough, more consistent, and more specific to the client's needs. They evidently see value in talking about and acting on the same things in the same way—in shared semantics and shared knowledge. The goal of the Syntelligence loan advisor is to get all the loan officers to base their loan decisions on the same knowledge. The Palladian Financial Advisor is a tool for getting all those in a large corporation who deal with financial information to mean the same thing when they refer to "profit" (for example), and to handle financial information (that is, apply financial knowledge) in the same way.

By using technology that shares knowledge throughout a company, we can improve productivity in two ways: some knowledge-based tasks can be automated (reducing the cost of labor), and some knowledge previously held by specialists in the company can be distributed to workers (reducing overhead). At the same time, other benefits, such as improved quality and improved timeliness, will accrue.

*Expect vertical-market knowledge bases to be built to serve as a platform for other products by the late 1990s or sooner.* If you want to hire someone to assist your bank's loan officers, you look for someone who knows about banking. If you want help in insurance risk assessment, you get someone who knows about insurance. When looking for an accountant, you prefer one who has experience in your business to one who hasn't, all other things being equal. By the mid to the late 1990s we will be seeing the basic knowledge of vertical markets encoded into systems which will then be used as a platform on which specific applications will be built. The knowledge platforms will serve as higher-level starting points for software developers (much as operating systems and user-interface toolkits do today for general programming), but they will be most valued for the degree of intercommunication between software systems that the shared semantics (the knowledge in the platform) makes possible.

Another important point about knowledge platforms is that they can be better designed for communication between knowledge bases. A banker who is responsible for international loans had better know about international economics in addition to banking; an agricultural banker had better know something of farming in addition to banking. With the kinds of knowledge platforms that have been described here, not only does the developer get a boost for developing a knowledge base for a particular field, but part of the task of integrating different knowledge bases into one functioning system will have been done for him—much as the UNIX operating system enables integration at the byte-stream level and the Apple Macintosh operating system enables integration at the level of cutting and pasting text, numbers, and pictures. Common knowledge bases will likely bring us integration at the knowledge level; we will be connecting concept streams or cutting and pasting ideas rather than text or pictures. Technically, this means that we will be able to move complex data structures between applications, and that the meanings of the elements of those data structures will be shared among the applications.

*Expect formal knowledge to be published in executable form, not just in textbooks, in the mid to late 1990s.* It will be possible to reformulate a significant fraction of the formal knowledge that is found in libraries and make it accessible for use by people who need not understand it (in exactly the same way that one uses a calculator to compute a square root without understanding the algorithm implemented on the calculator). Much formal knowledge is not used today because it is too difficult to learn how to apply it. Linear programming, statistical inference, queuing theory, and Markov chains would be more widely applied if they were more accessible. A first example of this is Palladian's Operations Advisor, which is based on queuing-theory models of factory operation. The queuing-theory basis has been available for some time; the contribution of the Operations Advisor is to allow those factory planners who recoil from the lambdas, rhos, and mus of queuing theory to benefit from the models without having to deal with the mathematics.

The number of systems developed will depend upon the cost and convenience of development. Currently it is rather difficult to develop extensive knowledge bases; thus, relatively few (thousands) have been built, relative to how many could be built. The difficulty of the programming is holding back the development of knowledge systems. If knowledge-base development were as easy as writing a book (which is not all that easy), one would expect the number of knowledge-based systems to be at least as large as the number of technical books. The market for these knowledge systems should be larger than the market for technical books, since the knowledge need not be understood to be applied. (Compare the market for books on numerical analysis against the market for calculators.) But today building a knowledge base is quite a bit more difficult than writing a book. The tools for developing the knowledge just aren't there. Today we primarily have tools for transcribing articulated knowledge into knowledge systems, not tools for helping one develop the knowledge. It is as in conventional programming environments: full screen editors, debuggers, and trace and break facilities all make it easier to develop programs on line, and fourth-generation languages and active displays (such as spreadsheets and Hypercard) make programming easier still. Making knowledge acquisition easier will entail eliminating mechanical obstructions, guiding the author toward appropriate structuring of the knowledge, and aiding him in dealing with the complexity of the task. It may also involve coordinating knowledge-base development among several knowledge engineers.

### Fundamental Limits of Knowledge-Based Systems

It appears that all the activities surrounding human knowledge (recall, learning, discovery, vision, integration of sensation, communication) could be duplicated on machines. It also appears that human capability does not represent any sort of limit. Each of these activities will probably one day be performed "better" in some sense by computers than it is currently performed by people—more quickly, more thoroughly, with greater precision, with greater sensitivity, with fewer errors. It will take some time, but will probably happen.

However, it is unlikely that computers will ever have the same knowledge that people have. There will certainly be considerable overlap, but much of human knowledge relates to the integration of a mind with a body. Much of what goes on in our heads relates to feelings and emotions that are there in part because of the way our minds and bodies are intertwined. Hemingway said that he wrote not to tell a story but to create the experience in the reader. If he reads a good paragraph about the moment a fish hits a line—recounting the quiver of the rod, the line going taut, and the whir of the reel—a reader who has fished recreates the experience and feels it happening again. We will probably never get this kind of understanding of the same paragraph in a computer. The computer will be able to get the facts straight, and even to infer some facts that have been omitted; however, it will never know the feeling of a rod quivering as a fish hits the line. It will know that the rod quivers, but will not know the feeling. The distinction is even more obvious in the case of a Stephen King novel: a computer might be able to recognize what parts would be scary, but it would never actually be terrified. In many applications, this distinction is insignificant or irrelevant. In some, however, it may be difficult for the computer to deduce what is important and salient in a body of text, a story, an argument for a new technology, or a political essay. These often have less to do with facts than with feelings.

*Expect knowledge systems to progress from "smart" to "brilliant" through the 1990s.* Today, knowledge is being captured and applied in knowledge systems. These "smart" systems are generally built to answer specific questions, often in highly controlled environments. Developments through the 1990s will enable knowledge systems to become smarter than "smart," in order to deal with more complex situations in terms of the amount of knowledge required and the uncertainty of the data.

The prime characteristic of brilliant systems ("brilliant with respect to what machines can currently do, not with respect to human capabilities) is their ability to manage their own missions. Whereas today's knowledge systems are just becoming integrated with existing conventional computing installations, a brilliant system will

• have direct connection to and control over varieties of sensors, integrating different kinds of sensor data into one coherent interpretation of its environment,

• use image understanding to combine the outputs of image processing (such as edges and histograms) with expectations to better describe what is in a scene and what is happening,

• understand its objectives, priorities, and constraints and plan its actions accordingly,

• replan its actions as events unfold if the existing plan becomes inappropriate to the current situation,

• collect a history of which strategies worked and which ones didn't, and analyze this (perhaps off line) to improve performance with over time,

• coordinate the actions of multiple cooperating knowledge systems, and

• combine numerical and symbolic processing to act fast enough to apply to "real-time" applications, such as factory control, interactive systems, and weapons systems.

The coming brilliant systems may well be used in factory automation. When most people think about factory automation, they think about robotics. Although robotics does indeed have a major role to play, we believe that information technologies have an even greater role in increasing manufacturing competitiveness. It is useful to think of this as a problem of command, control, and communications, with an object-oriented data base forming the hub around which the following tools are integrated: process planning, process control, process simulation and modeling, just-in-time training, maintenance and diagnostic expert systems, dynamic scheduling, inventory management, and multi-agent planning.

## Hypermedia

It is sometimes just as hard to imagine life without certain amenities as it is to imagine how much more convenient life would be with just the right improvements. The encyclopedia, originally conceived as a compendium of all that an educated person should know, has accumulated a number of conveniences over the years: cross-references, indexes, keywords and subject headings in bold type, index words listed on the top of each page, and (fairly recently) a tendency away from expansive descriptions of general subject areas and toward relatively short articles on specific subject areas arranged alphabetically rather than by subject (the *Encyclopaedia Britannica* uses both approaches). The encyclopedia of today is arranged for convenient browsing, not to be read from the first page to the last. It also typically has pictures, diagrams, graphs, and sometimes color overlays. An encyclopedia without these features would certainly be harder to use than what we have today, yet for years we have been working with computer files that have none of them.

Hypertext, as developed since the 1960s, puts cross-reference links in text files so that a user can jump from one part of the text to another. These links are used to relate an idea in a section of text to an example, to more general notions, to similar notions, to elaborations, to the history of the idea, to the biography of the author of the idea—whatever relationships the hypertext author chose to represent. Hypermedia is the same sort of thing, but in addition to text it involves graphics, video, sound effects, music, speech, animation, simulation—whatever related information might be accessible through a computer. In addition to these built-in links, a user browsing through a hypermedia document can build his own links representing his personal trails through the landscape of information.

The essential contribution of hypermedia is to make information more accessible by introducing and facilitating the notion of browsing—a notion different from the familiar computing themes of calculating, querying, modeling, and simulating. An example is Apple's uncommonly successful Hypercard system, which incorporates many of these ideas in a broadly used system but tends to project the notion of browsing as something separate and apart from other computing themes.

### A Possible Hypermedia Example

Imagine a volume of information available through cable TV for sports enthusiasts. While waiting for a game to start, one could do one's own pre-game analysis on what to expect. One could begin with an in-depth study of the effects of injuries at every position on both teams, based on a statistical analysis covering the last five years. To assess morale and determine whether the teams have "come to play," one could listen to this week's interviews of team members and decide whether one team's determination to "give 110 percent" is more convincing than the other team's resolution to "put on their game faces" and "give it everything they've got." For a comparison of game plans and strategies, one could call up John Madden's careful diagrams of plays made and busted. After this superficial survey, one could examine x-ray photos of hip pointers, hamstring pulls, and Achilles tendon injuries, plus motion pictures (fluoroscopy) of quarterbacks' wrists, thumbs, and rotator cuffs in action. And critically important, of course, would be the psychologists' assessments of each player's state of mind regarding press coverage and contract negotiations.

There is a lot of distance between where the technology is today and the delivery of this kind of information—particularly in the realm of making up-to-date information of this kind available in the home of the sports enthusiast on demand. But the capabilities of delivering information in a variety of forms, linking team information to player information to injury information, and blending text, numbers, analysis, images, motion, sound, and animation in one system are emerging. As storage and communications technologies continue to make the availability of images, sound, video, etc. more cost-effective, hypermedia technologies will play an increasingly important role in organizing information and facilitating access to it.

### Grain Size: From Knowledge Systems to Hypermedia

The concept of hypermedia is related to knowledge systems. They are two ranges on the same spectrum. Both deal with structuring information, and both have ways of traversing the structure. Knowledge systems focus on structuring small bits and automating the structure traversal. That's what inference is. Hypermedia focuses on larger pieces, the structure typically is handcrafted, and the traversal typically is done manually by the user, browsing through the knowledge space. Each of these extreme views misses important characteristics of the other. Where possible, hypermedia systems should rely on automatic indexing, automatic linking, and inference. At the same time, it would be a great mistake to consider only small-grain knowledge systems and ignore the vast amounts of electronic material becoming available as unstructured text, graphics, video, and so on. The important consideration is not knowledge systems or hypermedia systems, but combining the benefits of these two approaches to structured information and knowledge. As was observed in the November 16, 1987 *Wall Street Journal,* "What most people need from their computers now isn't another application but interactive information." That is the direction in which both hypermedia and knowledge systems are taking us.

### The Merging of Computing, Communications, and Images

Three technology areas that have had a profound effect on society are computing, communications, and images (the combination of print and video). These technology areas have all been steadily expanding. Today there is considerable overlap between them. In fact, the technologies are merging. We will be seeing more products that are clearly within the computer-communication-image realm, but which can't be classified simply as a computer product or a communications product or an image product.

This merging is shaking things up. Impediments to which we have resigned ourselves are being smoothed, and altogether new products and services are becoming possible.

Education and training place a high value on graphics, video, and sound, but the costs of authoring are a real barrier. Hypermedia can provide the flexibility and versatility to create multimedia training and diagnostic systems. Hypermedia could be the key that will cause optical disks to gain the position that was forecast for them. People live in a world of color TV and movies but are forced to learn, for the most part, from the relatively sterile environment of books and blackboards. This is likely to change.

In the management of large enterprises, hypermedia will provide a more efficient tool for simultaneous access to design, scheduling, inventory, production, and sales information. Current accounting systems do not manage or track information very well when they are forced to cross boundaries in non-steady-state, near-real-time situations. Hypermedia will enable users to cross data hierarchies more easily. This will have a profound impact on organizational hierarchies that are based on access to information. Organizations will become more horizontal, permitting more distributed decision making and thus coupling the "eyeball" and the "trigger" and making the old concept of "the boss" obsolete.

Consider the problem of spare parts for the military. If you need to find a part and you know the part's name or number, today's technology will serve you well. But if you have the part but not its name or its number, you are going to have trouble locating it in the inventory.

Language sometimes fails us. Trying to tell someone over the phone which hand is his right hand is difficult, but if you can point to his right hand the job becomes trivial. The merging of computing and images is creating the opportunity to combine the traditional advantages of databases in the handling of specific features (such as a product's weight, price, material, and supplier and the number on hand) with the benefits of images (shape, position, and relationships), allowing us to do image queries. Hypermedia systems are also making contributions in the combination of search techniques—for example, doing a query (as with a database system) or a classification (as with a knowledge system) to get in the right ballpark, then browsing (as with a hypermedia system) to explore possibilities and make final selections.

Confusing narratives on how to service a machine can be replaced with video demonstrations embedded in electronic repair manuals. Add to that a national wideband communication network (expected in the mid to late 1990s) and it becomes feasible for all manufacturers of consumer products to put diagnostic systems on their products into a central library, accessible from anywhere at any time. Before you take your car to a repair shop, you will be able to consult the manufacturer's experts (through expert systems) and get a good idea of what the problem is, what is involved in repairing it, and what it is likely to cost. By combining the benefits of wideband communications, repair manuals, databases, microfiches, and training videos, companies and consumers will solve familiar, nagging problems.

Another example of the convergence of these technologies is the electronic shopping mall. How can Sears describe and organize 6000 items in a computer database so that home shoppers can use it and feel confident buying from it? One approach is to recreate the image and experience of physically going through a store with departments, aisles, and shelves, and making the service available over cable TV. For example, the customer will find men's sweaters by electronically walking to the men's department, will see and hear about the sweaters, and will compare, say, the prices and materials of the sweaters he is interested in. The electronic shopping mall implements the notion of catalog sales with richer images from video, sound, and speech and with the computational benefits of query, comparison, and calculation. Other examples: the virtual university (outstanding lectures on many subjects would be available without leaving one's home); the virtual market (buyers and sellers for almost any commodity could be matched on the basis of features, price, and delivery time); the virtual factory (designers could assess the cost of having their designs produced by various manufacturers).

Systems that implement small versions of the kind of products we are discussing here are appearing now on PCs (especially on Apple Macintosh computers, with their phenomenally successful Hypercard hypermedia system). But there is a problem of scaling up to the size of the Army's spare-parts inventory, or to the size of Sears' product lines. Current limitations notwithstanding, the biggest surprise would be if we didn't see some astonishing and unexpected new products and services.

*Entertainment will be a major application of hypermedia in the 1990s.* The first commercially successful application of the notion of hypermedia (actually, there was just one medium—text—so *hypertext* is more accurate), to adventure games, was very successful. These games provided the computer adventurer with elaborate labyrinths strewn with treasures and fraught with mystery and danger. The adventurer would tend to get lost in the labyrinth almost immediately, which was all part of the fun. (One of the great challenges of hypermedia systems today is making them extensive without making them confusing and disorienting.) There was even an aftermarket for maps of labyrinths to help the adventurers get around and secure the treasures. Adventure games provide a fairly rich form of "out-of-body" experience, much as books, plays, movies, and radio stories do, but adventure games require a degree of personal control and participation. They are similar in control and participation to video games, another established entertainment medium. Video games have been limited to visceral experiences—racing, fighting, shooting—because they fit the current restrictions of graphics and of learnability in an arcade.

But if you haven't taken a good look at what's in the arcades lately, do it. The graphics are really impressive, and the games are more complex and interesting.

There is an aspect of computer games that hasn't been talked about much in recent years. Late one night in 1974 one of the present authors, Tennant, walked into a terminal room for the then-experimental PLATO CAI system at the University of Illinois, thinking he would have the place to himself and get some work done. No way. The place was packed. All were intensely absorbed in their screens, occasionally muttering "damn, damn, damn" under their breath or bursting out with "All right!" and smiling and shaking their fists in the air. After 10 P.M., it turned out, Empire players were allowed on the system. Empire was a computer game, but it was not like those we have become used to. The excitement and energy in this room was generated by, not just man against computer, but man against—and man in league with—man, the game being mediated through the computer. These people were playing a space battle game with other people all over the world, connected through PLATO's educational network. To play, you chose what kind of person you wanted to be; then there you were on the screen amid the others. If you weren't destroyed in the first few seconds (Tennant usually was), you could contact other individuals, send secret messages, make pacts, conduct coordinated attacks, double-cross your partners, or fly off in another direction—where you might encounter another bunch of people vying for supremacy in that portion of the universe. Tennant resisted the temptation to become an Empire junkie himself, but the atmosphere in that room was electrifying.

The Defense Advanced Research Projects Agency and the U.S. Army collaborated in building SIMNET, a system that enables tank crews to conduct simulated engagements with one another in geographically dispersed locations through the power of lifelike graphics and networking. In the future, the parameters and the worth of weapon systems might be established by testing them in simulated engagements involving potential operators.

When motion pictures were new, it was thought that their main applications would be training films and travelogues—until the first movie for commercial entertainment, *The Great Train Robbery,* came along. Training films and travelogues do get made, but today movies are primarily *movies*. There may be substantial markets for electronic shopping, but the technology can also let you pretend to be a pilot in an authentic reconstruction of the Battle of Midway, flying wingtip to wingtip with your cousin across town—or trying to find treasures in the passageways of the rusting hull of the *Titanic* at the bottom of the Atlantic, avoiding Jaws and a giant squid and competing against and cooperating with your neighbors—or conducting sensitive diplomatic negotiations with friends and strangers from all over town to advance the cause of your imaginary nation. It's the same technology—which would you rather do?

### Support for Information Flow

The value of information and knowledge, in whatever form it takes, is often a function of time. The information in today's newspaper sells for a quarter, but yesterday's newspaper has only the value of the paper it is printed on.

The flow of information around us is often the most valuable information to us—often more valuable, bit per bit, than the much larger volume of collected information. Management information systems are capable of collecting huge amounts of data, yet output in useful form is too often "history." Flowing information can be the hardest information to make use of, especially automatically. Generally, the more unexpected and the more surprising the new information is, the more valuable it is. (In fact, this is reflected in the way information is defined in the realm of information theory: the amount of information in a message is a measure of the amount of surprise in the message.) The more unexpected the messages are, the harder it is to anticipate them, so the harder it is to prepare to deal with them automatically—but the more valuable it is to do so. We believe that there are prospects for automated systems capable of extracting information from data in a timely manner. These systems will be capable of pursuing user-directed goals over time.

The flow of electronic information has taken firm hold in some organizations, particularly in the form of electronic mail. To many people, it is a communication medium ranking in importance with the telephone. It brings in not only individually addressed mail messages and local discussion but also information available over the nationwide networks, such as discussions of politics, science fiction, artificial intelligence, superconductivity, movie reviews, and computer interfaces. Other kinds of flowing electronic information are newswire services, stock quotes, videotext, weather information, closed-captioned transcripts of television programs, and (in analog form) television and radio broadcasts themselves.

Electronic information is still relatively primitive. For example, wire-service news copy resembles reporters' drafts (which in many cases it is) more closely than it resembles a published newspaper that has been edited and carefully laid out. It bears little similarity to a glossy magazine.

As computers become connected, the volume of electronic mail will increase (although progress here could be impeded by a lack of communication standards). A more widespread user base will tend to justify and demand improving the quality and variety of what is available. But the most important changes will be qualitative: the flow of information will be, to some extent, understood and acted upon by machines.

A prerequisite to understanding is structure. The more articulated the structure of a message, and the more specific that structure is to the content of the message, the more action can be taken automatically. Two ways to give structure to messages are for explicit structure to be provided by the author and for derived structure to be determined by a language-understanding program.

### Natural-Language Processing

Natural-language processing is an area of knowledge technology that has long been studied in hopes of commercial application. Although there have been commercial applications of language translation, natural-language interfaces to database systems, and text-understanding systems, none of these have been broadly applied.

The strongest potential seems to lie in text-understanding systems. The main problem here is coverage, and it is approached in two ways. The first is to build only very narrow applications. One example of a system in regular use is the AMVERS system developed for the US Coast Guard. It reads messages from merchant ships on the open seas reporting their identification, location, bearing, destination, and possibly other information. The information is put into a database that tracks several thousand ships at sea worldwide so that, in the event of an emergency, a nearby ship can be called upon to render assistance. The position reporting is voluntary, and the messages tend to be undisciplined. Previous attempts to get the ships to use fixed message formats failed, but the AMVERS text-understanding system for this highly restricted domain has been successful at automating the encoding of about 80 percent of the messages—enough to pay for the requisite development and machines in two years.

The other approach to the coverage problem for text understanding is to choose applications where incomplete understanding is acceptable. *Data extraction* is a better term for this than *understanding*. These systems are designed to pull certain facts from text. For example, a data-extraction system could scan the *Wall Street Journal* for facts about a company and its competitors—product announcements, earnings reports, plant construction starts, layoff announcements, changes in management, and so on. These facts could then be recorded in a database, which could be used to ascertain the current state of a market. Most applications of this kind have been in the intelligence community, often under security restrictions. However, this technology could have a considerable impact on our ability to automate the use of the flow of information.

*Expect a breakthrough in text processing by the mid to late 1990s.* Since about 1970, a lot of attention in the natural-language-processing research community has been focused on more complete grammars and better parsers. These grammars are now quite complete for English, but the coverage—the range of concepts that a natural-language system can understand—is still so limited as to make natural-language-processing applications rare. A possible avenue to providing the broad coverage required by language understanding is through various dictionary projects. Machine-readable dictionaries are a by-product of dictionary publishers' move to computerized typesetting. Now they are being used by many computational linguists for a variety of projects. The definitions of words is a long way from the knowledge of the world that humans have. However, the large machine-readable vocabularies (some in excess of 300,000 words, and some with shallow semantic definitions rather than just text strings) are so much larger than the vocabularies previously used in natural-language understanding (fewer than 10,000 words, and usually fewer than 3,000) that we should not be shocked if a radical shift in natural-language technology occurred by the mid 1990s. The most useful products will probably be related to data extraction from text rather than to user interfaces, for three reasons:

• Other technologies already available work better than even ideal natural-language interfaces.

• Data extraction does not require full understanding of the text; partial success is still useful.

• Text will always be with us, and will always need to be read and understood—there will be no substitute technology, as there was for user interfaces.

### Providing Structure through Forms

If text cannot be structured automatically by language-processing technology, we can still make use of structure provided by the user. The structure can be indicated through the use of forms—different forms for different kinds of messages. Not only is the structure itself interpretable, but since the contents of a slot in a form are restricted, it is easier to build understanding programs specific to each slot to determine the meaning of what has been entered. Communication through forms is being applied to electronic mail. It promises to automate part of the knowledge worker's task of communication, which accounts for a significant portion of a knowledge worker's time. (One of the most interesting and promising projects of this kind is in progress at MIT: the Information Lens project under the direction of Tom Malone.) Many benefits can come from simple form-based structuring:

*Sorting*   Incoming messages can be sorted into bins by sender, by subject (e.g., all weekly reports in one bin), by number of addressees (junk mail), or by sender status (messages from my boss or his boss or my wife). Having to deal with mail through a linear sequence of messages is opaque.

*Busy/No Answer*   I would like to set up special relationships with some people; for example, if I send a message to my secretary and she isn't there, I'd like my mailer to tell me that immediately. That way, I will get some of the immediate feedback that a telephone busy signal or unanswered ring gives.

***Ticklers***   The mailer would be a good basis for managing events that stretch over time. I would like the option for my calendar to send me tickler messages, saying, e.g., "You'd better start on the presentation to Amalgamated, you have to give it in a week" or "It's time to check in with Felix to see how the mailer project is going."

***Structured Messages***   Next, messages can be given more structure. The most elementary example is an indication that a particular message requires a response or other action by the recipient—it could go into an action bin, without the need to understand the contents. More elaborate structure allows some messages to be composed or read as forms. This enables a level of message automation; for example, consolidating weekly reports from branch members into one branch report could be done automatically for the first pass.

***Reactive Messages***   Messages could combine graphics and speech with text. They could be made active, so that part of a message could include a spreadsheet that could be manipulated by the reader. Messages could be animated, behaving more like TV commercials than like telegrams.

***Emissary Messages***   Messages could be given sufficient structure and mail systems sufficient interpretation capabilities so that the message would be readable by both humans and machines. Hence, if you were to send me a message about a meeting on Thursday, my system might enter the meeting on my calendar, or it might reply that I am going to be out of town (except if the message warranted a change in my plans, in which case my system would attempt to locate me to let me know immediately).

   As more transactions become interpretable by machine, personal knowledge bases can accumulate. (We already have a lot of accumulated information in our file system, but today it isn't structured in such a way that it can be used automatically.) This allows the possibility of *emissary messages,* which not only carry a message but can answer questions to clarify the message by referring back to our accumulated personal information. If we get the message "Party at my house after the meeting—Sam," we may have some questions. Where does Sam live? What meeting? Who is Sam? Who else will be there? If a knowledge base has been accumulating in Sam's personal environment as a by-product of his daily activities, there may well be answers to these questions and more that Sam's environment can provide. An emissary message will alllow me to ask the questions without bothering or embarrassing Sam, and to make use of the context of the message (which meeting, which party, which Sam).

Some of these questions may be asked automatically by our environment when the message is received. (For example, we have stated a goal of meeting someone with a hot-air balloon.) In principle, all these things are fairly straightforward. They are not in common use because mailers either exist on time-shared mainframes or rely on lowest-common-denominator terminals. Now that PCs as terminals are becoming the norm, sharing the processing load between the PC and the time-shared machine will be figured out soon. And hypermedia-related ideas of structured documents and integrated applications are becoming more commonplace. This, coupled with the importance and the opportunity of improving the communication of knowledge workers, suggests a high likelihood that electronic mail systems will soon improve dramatically.

### Conclusion

We have tried to limit this discussion to "surprise-free" projections, meaning that we wouldn't be surprised if any of the things mentioned here actually came to pass. Much more likely is that some unforeseen or overlooked technology will burst on the scene, changing the ground rules and the opportunities, just as transistors, integrated circuits, microcomputers, spreadsheets, direct-manipulation user interfaces, and Apple's Hypercard have done. After all, as Niels Bohr once said, "Predictions are always difficult, especially about the future." The value of looking ahead is to set aside some time to imagine what our opportunities might be, and to prepare ourselves to evaluate those unanticipated developments when they do come (as they most certainly will). Our time to react is limited, and opportunities may be missed. Niels Bohr notwithstanding, it's easier to predict the future than to change the past.

# 5

## Anything, Anytime, Anywhere: The Future of Networking

*William R. Johnson, Jr.*

The future of networking, like the future in general, will not appear suddenly, in sharp contrast to a darker past. It will unfold before our eyes. We all remember the world before ATMs and bar-code scanners. But can we pinpoint when the "future" happened? Was it the first time we noticed the terminal offering 24-hour banking? The first time we encountered a supermarket with a new kind of computerized checkout? Or did the future happen when these things ceased being novelties and became an everyday part of our lives?

The future of networking can already be seen in outline in new ideas, new services, and new capabilities.

### Anything, Anytime, Anywhere

In the 1990s, networking will evolve to the point where, for all practical purposes, people will be able to electronically communicate anything (voice, data, image, video), anytime, anywhere in the world.

Predicting the future of networking is simple. Answering the question "Who will benefit from networking—and how?" is much more complex, but clearly the chief benefactors will be persons who are involved in integrated work processes that span the enterprise. Determining how best to employ networking will become an increasingly crucial part of every business operation. Today, three aspects of networking promise to make learning how best to use it a very complicated, very important, and very exciting undertaking.

First, it's what you don't see that makes a network work. An automatic teller machine connects to a network of computers that transfer money from one account to another. A bar-code scanner in a supermarket is part of a network that sends purchase information to people in distribution, sales, marketing, and management. In both cases, behind the scenes, a complex diversity of devices, connections, signals, protocols, and applications are working together to present an illusion of simplicity to the people using the network. This invisible technical infrastructure can be referred to as the "network utility."

Second, computerized networking is an evolutionary step in human communication, one destined to have more impact on our lives than television or the telephone. Networking will increasingly transform, multiply, and concentrate human interaction. It will change how we relate to one another, how we learn, how we define and reward work, even how we govern ourselves. Just as the proliferation of ATMs has changed our daily routines, our image of a bank, and even, subtly, our ideas about money, networks will increasingly transform our notion of wealth and credit. Just as the scanner has changed the job of the cashier and the store manager, and, fundamentally, the relationship of the store with its suppliers and headquarters, networks will increasingly transform our definition of work and the relationships we have with co-workers, managers, customers, and suppliers. These changes can be referred to as the "social" aspect of networking.

Finally, networking, by its very nature, is not a solitary or linear endeavor. A network is created not only by what you do but also by what others do—and by what you do in response to others, and what others do in response to you. The real promise and the real challenge of networking come from the exponential possibilities of the myriad interactions and combinations of interactions it makes possible. We can call this characteristic the "relativity" of networking.

### The Future of the Network Utility

Think about picking up the phone to call a friend or scanning through information from another computer on your screen. When the network works, you are unaware of the transmission media, the routing algorithms, and the software being used, and you don't think about where the database is located. In fact, "not having to think about it" on the user level is one of the key criteria of a network utility.

But in order for a network to work, someone has to think about it, be responsible for it, and manage it. Two of the biggest questions organizations will face in the 1990s will concern what investments they should make and what kinds of expertise they should develop. That is, what part of the network do we want to own and control and what part do we want to buy from others? The answers to these questions are inextricably entwined with a firm's overall business strategy (the chain of added value that determines organizational success). They cannot be delegated. And as technology and market conditions change they will have to be revisited. The managers who are going to be successful in the next decade are those who understand both how the potential of networking expands their strategic options and how to evaluate the technical, organizational, and economic tradeoffs in implementing networks.

To perform as a utility—that is, to usefully communicate anything, anytime, anywhere—a network must have the following four characteristics:

- connectivity
- interoperability
- manageability
- distributed applications and connective services.

Connectivity is the ability to move information from point A to point B, regardless of the diversity of media (wires, fiber, air, vacuum) or transmission technologies involved. It is the "anything to anywhere" part of the networking utility. Often, because it is more tangible than the other requirements for networking and because it is the first requirement to be tackled in building a network, connectivity gets the lion's share of attention. In the 1990s, however, many of the challenges of connectivity will be met, and the spotlight will move to interoperability, manageability, and distributed applications. Nevertheless, connectivity is the first prerequisite for a networking utility. To send "anything" requires a "pipeline" with adequate capacity and speed. To send it "anywhere" requires the ability to interconnect with other networks around the world.

### The Physical Pipeline
Today data, voice, image, and even full-motion video can be digitally encoded and sent across a variety of physical media: wires, fiber optics, microwaves, and satellites. Planning can now be based on the assumption that all information will be digital.

The greater the number of systems, people, and processes that want to transmit information to one another, the greater the speeds and bandwidths required. Much as with water in a pipe, to get more information through you can increase the rate of flow (speed) and/or increase the amount that can flow through in a given unit of time (bandwidth).

New applications under development today presage tomorrow's need to transmit very large quantities of data very quickly. These include the following:

high-quality digital interactive video transmission

computers that understand, capture, and recreate voice

hand-held intelligent receivers that pick up satellite signals relaying exact position, maps of the surrounding territory, and suggested routes to desired destinations

optical scanners that convert text and graphics into digital forms

x-ray images sent to specialists in other cities for help with a diagnosis.

As computing costs continue to drop, computing power and intelligence will continue to be increasingly distributed in many locations—and increasingly portable. People will be able to do more things electronically from more places, to share their results with more people more quickly, and to represent their ideas in many ways. Eventually, people will expect the convenience of plugging into "the network" wherever they are. The demand for more speed, greater bandwidths, and integration of voice, data, and image on a single medium will continue.

Technology will evolve to support that demand—and to inspire new applications. For a number of years Ethernet has been providing useful networking with a speed of 10 million bits per second. Today, fiber optic technology, with a speed of 100 million bits per second, is beginning to be implemented. Continuing this pattern of tenfold improvements, the next development promises 1 billion bits (1 gigabit) per second. As amazing as it might seem, I believe gigabit technology will be needed to support comparable improvements in processing and applications before the end of the century.

Obviously, it would seem, the bigger and faster the pipeline, the better. But gigabit technology does not exist yet—and if it did, the cost would prohibit its use for many applications. More important, no applications existing today require this speed, and it is likely that the majority of organizations will not have applications that require it for many years. However, a company charting its networking strategy for the 1990s should be aware that such a capability will eventually be feasible. The practical questions concern not only what we need to do today and what is the most cost-effective way to do it, but also what we will need to be able to do tomorrow and how we can best incorporate new technology when it becomes available or cost-effective.

A group of design and manufacturing engineers performing simultaneous, distributed work by interactively sharing complex product and process designs will need a more robust pipeline than a franchise sending a rollup of the day's business transactions to headquarters. But how should pipeline requirements be sized over time? Organizations need to consider future use, equipment, and service costs, as well as tradeoffs in quality, time, and labor. For example, sending a photocopy of a page of information by express mail is more cost-effective than going out and buying a fax machine to send a facsimile of the same information. However, if increased use is projected, and time and labor are factored in, a fax machine may soon be cost-justified. If, on the other hand, the person to whom you want to send information doesn't have a fax machine, the cost savings means nothing. Similarly, the return on an expensive high-speed pipeline may not develop, because it has to feed into a slower public link.

It soon becomes evident that flexibility and the ability to grow without jettisoning significant investments are more powerful advantages than raw performance. To respond quickly and easily to changes in business or technology requires investing in technologies that are both reliable and flexible. The risk of investing in an unstable or obsolete pipeline technology can be minimized by choosing a data-link technology that supports useful networking over a wide range of media (twisted pair, baseband and broadband coaxial, thin coaxial, satellite, microwave, fiber optics, leased lines, etc.) and that is supported by a large number of vendors. The flexibility to incorporate a new pipeline technology in the future, without affecting existing investments in network applications and devices, can be achieved today by choosing a networking architecture that provides independence of function for physical media, routing, protocol, and application technologies.

### Rent, Lease, or Buy?

To be able to send information anywhere, a network has to be able to connect physically with a wide variety of devices, and to interconnect with other networks. Very few enterprises will find it feasible or desirable to build their own global network; most will, at some point, plug into the networks of others.

The 1990s will certainly resolve any remaining technical gaps in physically connecting devices and networks. Connectivity decisions will become economic decisions.

The decision to install or buy service will vary from company to company. Tradeoffs between owning and leasing will have to be constantly reevaluated in light of changing requirements and economic benefits. A small company may install its own data-communications network on the premises and buy outside services for data processing and for all its voice communications. However, frequent communications with a supplier may justify installation of a private data link between the two companies, and as the company grows it may save money on installing its own telephone branch exchange.

Integrating voice and data over the same physical medium using the emerging Integrated Services Digital Network (ISDN) standards will soon enable voice carriers to provide data-communication services at speeds of 56,000 or 64,000 bits per second, in addition to voice communication services on the same physical wire, thus adding to the connectivity options available to companies.

Some people say that with ISDN "the data rides for free." This is because in most companies today 80 percent of the traffic is voice; the other 20 percent is data. This allocation is moving toward 50/50, however, and by the end of the 1990s it may even have reversed. If this happens, it is important to remember that the requirements for supporting distributed data communications are very stringent. When processing capabilities, data storage, software storage, and peripheral-device access are spread across remote locations, data transfer must be swift. In today's environment this means speeds in the megabits-per-second range. I expect ISDN to support speeds of 2 megabits per second by the end of the 1990s. However, today's basic rates for service to each desktop will not adequately support the data-transmission rates required for distributed computing services. In addition, it is likely that higher-performance wide-area transmission capabilities will evolve by degrees and by geography, becoming available in metropolitan areas before they are available on a national or international basis. Local-area network media will continue to deliver data-communications services to the corporate desktop for some time.

With the divestiture of AT&T, some large corporate users have discovered that "being your own Bell" pays off. In addition to providing services to their own users, they are able to sell excess capacity. As Tom Valovic points out, "this . . . phenomenon has now moved the telecommunications department from a liability to an asset in the corporate balance sheet."[1] It also moves private companies into the business of providing public network service. Again, because connectivity is one of the more tangible aspects of networking, it is here that regulation issues will first surface and will first be resolved.

### Interoperability

In much the same way as connectivity describes the physical integration that must be in place to support the network utility, interoperability describes the logical integration that must be in place. Interoperability is the ability of diverse intelligent devices to communicate with one another in performing meaningful tasks. A network that provides connectivity without interoperability provides the "plumbing" to communicate anything, anywhere, anytime, but not the intelligence to do so.

A study conducted in 1986 found that more than 90 percent of existing computer networks used equipment from two or more vendors—and more than 50 percent relied on at least five separate vendors.[2] The variety of devices in organizational networks will increase in the 1990s as networks become larger, telecommunications and data communications converge, and users insist on protecting both past investments and future options.

Even relying on equipment from one vendor does not solve the problem of interoperability. Most major computer vendors sell systems as diverse as if they came from different suppliers. In any case, I don't believe—and neither do most users—that any one vendor can meet all current and future networking needs. As the MIS manager at a large corporation remarked to me recently, "You don't want to have to deal with a hundred vendors in your shop, but you don't want just one either."

Today there is a thriving industry in interoperability products—software, or the so-called black box, that translates between systems that can't talk to one another. This industry will continue to prosper through the 1990s. But this approach adds cost and complexity to networks—and, worse, it creates an atmosphere of unpredictability that threatens the whole industry with chaos or paralysis.

### Open International Standards

The issue of interoperability is so crucial to everyone's success that major players in the industry—vendors, customers, user groups, and government agencies—have come together to support the development and adoption of common standards.

To date, the International Standard Organization's Open Systems Interconnection (OSI) standard has been accepted by more than 100 computer, communications, and software vendors. The OSI model defines a layered architecture that specifies standards for seven aspects of interoperability: physical, data link, network, transport, session, presentation, and application. This OSI architecture also provides for "logical independence" between these layers, so changes in one function can be made without affecting other functions.

Together with the emerging ISDN standards, OSI provides a basis for the interoperability of local-area networks, wide-area networks, public and private networks, and voice and data networks. It will be the mechanism for improving interoperability in networking products for the foreseeable future.

Paradoxically, it is adherence to standards that will preserve diversity, freedom of choice, and innovation in the future of networking. Standards define what is expected and what is allowed, enabling smaller companies to enter the market and larger companies to focus their unique strengths in creating new products and services. When all the railroads in the United States had different-sized tracks, "innovation" was determining which width of track would be best. Only when track gauge became standardized were resources applied to developing bigger, more powerful engines to pull more cars. In the same way, as a result of OSI, the next ten years will see an explosion of creativity from independent software vendors designing distributed applications to play on the networking "platforms" developed by other companies.

Standards for interoperability (versus standardization on one particular product or product line) allow all vendors to specialize in what they do best and then to integrate with the rest, rather than trying to develop their own homogenous, "winner-take-all" solution. Standards of interoperability also give users the freedom to change vendors and technologies, allowing them to buy the best solutions for their needs with confidence that they will work together.

Standards will also accelerate the rate at which organizations adopt new technologies. To improve means making changes; however, when making a change means ripping everything out and starting over, improvements come very slowly. The layered architecture and the ongoing development of OSI provide for a nondisruptive, predictable way to incorporate new technology into OSI-compliant networks. Because one can take advantage of new technologies while protecting one's existing investments and operations, one can afford to experiment with improvements, learning as one goes. This "bottom-up" approach allows companies to implement technology in small steps, evaluating and making corrections along the way. It is no longer a question of "betting your job" or "betting the company" on a "once and for all" decision. A standard architecture lets you start as small as you want and grow from there.

### Interoperability between Networks

As networks grow, interoperability between networks becomes as important as interoperability between diverse systems. Local-area networks will be the building blocks for creating larger, more geographically dispersed networks. Analysts agree that local-area networks will continue to provide the dominant transport in the corporate network through the 1990s. Gateways and bridges between similar and different local-area networks will become more sophisticated in delivering network interoperability, both within buildings and across geographies. The future will require greater congruence between local-area and wide-area networking technologies to allow a modular approach to the building of extended networks.

Interoperability in a multi-vendor environment is the best way to maintain diversity and choice for business networking in the 1990s. It will be achieved by adoption of and conformance to the OSI standard protocol model. By establishing the "playing rules," standards like OSI will allow the game to continue.

## Manageability

The third requirement of a network is manageability. Once a network is connected and interoperable, it has to be easy to manage and to monitor. It must also be open to new applications. And it must provide secure, uninterrupted service. Changes and enhancements to the network must be easy to accomplish without disrupting operation.

In the 1990s computer and telecommunications vendors and users will find network management a growing concern as the number of computers and applications increases; as voice, data, image, and video are supported; and as networks operate on a global scale.

To date there have been two basic styles of computer networking: a *hierarchical style*, in which one computer is designated as the "manager" and oversees the use of the network by other computers, and a *peer-to-peer style*, in which every computer "manages" itself. In the first case, devices "ask for permission to speak" and are then directed by the managing computer. In the second, devices are all of equal status and there is no prescribed hierarchy for communication.

In an environment of rapid change and high complexity, the peer-to-peer approach has been shown to be more advantageous than the hierarchical style of networking. By establishing "rules of the road" and relying on the combined intelligence of all members of the network, peer-to-peer networks are able to cooperatively and automatically determine the best routes— and to reroute around problems.

Like a human body maintaining its temperature, breathing, equilibrium, and so forth, normal operation in a peer-to-peer network is not directed through conscious effort by the "head." It is, instead, the result of continuous sensing, feedback, and adjustment by all the systems and subsystems communicating within it. The ability of each device to "feed forward" its intent and "feed back" its status simplifies network management greatly. In the same way that a self-regulating body frees the "head" to think about where it is going, to make changes, and to respond to unusual situations, a peer-to-peer network allows network managers to focus on more strategic issues.

In the 1990s it will be necessary to manage networks with a million users. It is generally accepted that networks of this size, complexity, and geographic distribution must be of the peer-to-peer style. Recent cost-of-ownership studies by Michael Treacy of the Sloan School of Management at MIT have found that comparing the total costs (i.e., the costs of hardware, software development, maintenance, support services, and operational staff) of even much smaller networks over time reveals cost differences of as much as 1000 and 1500 percent between hierarchical and peer-to-peer networks.[3] The next few years will see hierarchical networks redesigned to perform more as "peer-to-peer" networks, but it is doubtful that these hybrids will catch up to the performance of true peer-to-peer networks.

In addition to being largely self-managing, networks must be easy to change. If a network provides connectivity and interoperability but makes it difficult to implement and manage change, it is useful only in situations where the business needs are static. As networks get larger, the ability to add new technology or to add or move users without bringing down any part of the network will become a must. For example, the Digital Equipment Corporation currently adds 100 to 200 new users to its worldwide computer network every week. Our network would not be much of a utility if we had to take it "off line" every time we wanted to add new systems.

### Who Manages the Network?

Even if a network is relatively easy to manage, someone must be responsible for its cost, its performance, and its growth. Here, again, the question facing organizations is: Where should the responsibility and the control for managing the network reside? With the users? With corporate MIS? With the telecommunications organization? With the vendor? Again, the answer will vary from organization to organization, and over time.

At Digital, we own and operate one of the largest private data networks in the world. It connects more than 75,000 users and 30,000 computers in more than 300 locations in 40 countries around the world, supporting more than 100 distributed net-

work applications. In the mid-1970s we had only small, uncon-
nected islands of networking, each using different computing
systems with different operating systems to support a few specialized applications. Throughout our networking history, the
question of who should manage the network has run the gamut
from being little more than an afterthought to being a hotly
contested issue. It might be worth a detour into the past to consider who will manage the network of the future.

The management of Digital's small early networks was *ad hoc*
and varied from group to group. However, as the networks
grew to support the needs of each department, such as manufacturing or finance, they generally began to be managed by a
departmental MIS manager.

With more systems in place, the desire to communicate horizontally across departmental boundaries grew—especially with
the development of an electronic mail application for sending
and receiving corporate memos. This demand coincided with
the development of local-area network (LAN) technology. As
local-area networks grew in number, some retained an organizational orientation but others provided service for a specific
geography. These networks were managed by very independent LAN managers, in a loose confederation.

Eventually, to support corporation-wide networking and take
advantage of economies of scale, all the LANs were plugged
into a single wide-area network. Many thought the next logical
step was to manage the whole network as a "corporate utility"
from one central location. But local user groups fought this
concept vehemently, feeling that the service and flexibility they
needed would be lost in a corporate bureaucracy. Both factions
were right. Managing a large network, it turns out, is like managing a large road system. Local managers are needed to understand and support local needs and to respond to changing
conditions. Central, unified management is needed to establish
the consistency and efficiency necessary for interstate express-
ways. You want the local roads and the interstates to work together, but you also want them to be managed appropriately.

Every local-area network is different. The physical sites are
different, the applications required by the users are different,
and the levels of service required are different. In addition,
usually about 80 percent of communication is local. To provide
good service from a remote, centralized organization is very
difficult, if not impossible. To understand what people are

trying to do and what they need—as well as the unique local peculiarities, constraints, and opportunities—requires a local manager.

On the other hand, managing the larger "backbone" utility that LANs plug into and taking advantage of economies of scale calls for a central organization. For example, if a company is going to need 200 circuits, having a single entity represent the company and negotiate a price is more cost-effective than having the manager of each LAN buy two or three circuits at a time.

Finally, both local and central management require highly skilled technicians whose expertise they can rely on for design, analysis, and maintenance. Today three separate and distinct groups interact and cooperate to manage Digital's network:

the corporate data network management group, which as part of the corporate telecommunications group provides and manages the wide-area network, known as the WAN backbone, that local area networks plug into,

the managers of the local-area networks, who define and meet specific local operational needs, sit on the corporate data network "board of directors" to help define their service requirements for the WAN backbone, and pay for the corporate networking utilities they use, and

a corporation-wide network services organization with local branch offices which provide the technical resources and expertise to support the LANs and the corporate WAN. They charge both the corporate data network management group and the LAN managers for their services.

We at Digital believe this approach to network management will serve us well into the foreseeable future. The 1990s will find many organizations making decisions about how to manage their networks. Different organizations will employ different strategies—for example, many companies will find it more cost-effective to purchase all or some of their network management from others. However, I believe most successful network management strategies will share common principles. These will include the following:

• checks and balances between local and corporate needs

• broad representation of user constituencies

• contractual agreement for services.

### Network Security

As networks grow and interconnect, security will grow even more important than it is today. Intense effort and activity will undoubtedly result in a broader range of security solutions.

Security presents difficult challenges. A secure network requires security as an integral part of each node and each link, and in the behavior of every person who uses the network. Every node will have to be able to know the identity of every other node and be able to be relatively certain of its identity. It must not be possible to disguise a node logically and pass it off as if it were some other node. There will be some method of authentication between nodes, some kind of "fingerprinting."

In addition to protecting against logical intrusion at the node level, networks must protect against physical tapping at the link level. For that level of protection, encryption and decryption capabilities will have to be built into each node.

Each individual using the network has to be responsible for the security of his part of the network. Each object, file, or program will have an associated "access-control list" that expressly lists who can have access to it. As with the contents of my file cabinet, I am responsible for maintaining the access-control list to the things I have created on the network. Likewise, I am responsible for asking permission to have access to other resources on the network that I need and getting myself added to the appropriate access-control lists. Managers of local networks are responsible for controlling access to the networking assets for which they are responsible.

It would be much easier, and seemingly more secure, to set up a single, central authority to control all access to the network, but that is like making everyone who gets on a subway enter at one station and file through one turnstile. It is too cumbersome to be practical. Such an approach would be like making each person in a corporation ask the CEO for permission every time he wanted to share some information in his cabinet with a co-worker. Out of necessity, people bypass policy bottlenecks, and such a network is secure in name only.

At first it would seem that a network cannot be too secure. But all security exacts a price in terms of inconvenience, inefficiency, and cost. The proper level of security will vary from one organization to another and from one part of an organization to another. Levels of security must be flexible enough to be able to be appropriately set throughout an organization. For some applications, very elaborate security will be necessary, involving frequently issued key cards, locked disks, and the like; however, in most cases such measures are not worth the added expense. In addition, it is important to keep in mind that there can never be a perfectly secure system. However, as new technologies such as voiceprint validation become affordable, it will be easier to approach the goal of maximum network security with minimal inconvenience to the authorized user or application.

The network provides the capability to communicate anything, anytime, anywhere, but that is a capability that must be planned for, managed, and controlled. The future will bring complex, dynamically changing and growing networks over which all manner of people and devices will have equal access. Such a vision will require a comparable vision of network management to support it.

### Distributed Applications and Services
Once a network is interconnected, interoperable, and manageable, it can perform useful work—if it has the distributed applications that provide easy-to-use access to tools, data, and resources across the organization.

The fourth aspect of the network that organizations must consider as they move into the 1990s is: Do the applications exist to do what we want to do over the network, and if they do not can they be developed?

### New Applications

The potential for new networking applications is exciting. The 1990s will see a proliferation of enterprise-wide applications, such as electronic mail, as well as the increased use of such applications between companies.

The electronic exchange of information *between trading partners,* called *electronic data interchange* (EDI), will allow us to transmit business correspondence (such as purchase orders, bills of lading, and shipping records) across a network instead of sending them physically through the mail. This will significantly increase the timeliness and the accuracy of information, and will reduce the costs of document creation, storage, and postage.

The concept of EDI is not new; in fact, almost all of the "*For-*

*tune* 100" companies today have some kind of electronic information interchange with their trading partners. However, most EDI applications today are one-of-a-kind, customized networks, based on terminal-to-computer connections, like those between travel agencies and airlines or between pharmacies and pharmaceutical companies.

Tomorrow's computer-to-computer EDI applications will make extensive use of distributed intelligence and processing power to improve competitive advantage. For example, the exchange of CAD/CAM files for part design, modification, and coordination between suppliers and customers will reduce inventory, improve quality, simplify documentation, and reduce time to market. Eventually, advanced networking and distributed database technologies will make possible parallel design of products and integration of services between trading partners.

With the increasing implementation of networking standards, smaller companies will begin to participate in EDI networks—not only with their large corporate customers, but with one another as well. Many new kinds of applications will be developed as these networks weave a finer fabric of communication. In addition to one-to-one communications (such as electronic mail), distributed applications will support one-to-many, many-to-one, and many-to-many communications between people and between companies.

Videotex is an example of a one-to-many communication application in use today. A menu-driven program, Videotex helps many users find the information they need from information provided by a single source—for example, the latest information on employee benefits from the corporate personnel department. In France, Videotex is used in a national information system providing more than 1500 information services from over 800 sources. Service is available through inexpensive terminals, called "Minitels," which can be purchased, leased, or used in public places (such as post offices). A smart card reader, attached to the terminal, identifies the user for such purposes as banking and shopping. A directory service lists all 23 million telephone users, and an electronic mailbox allows users to reply to service providers.[4]

A many-to-one communications capability allows many people to input to single source, for, say, electronic surveys or polling.

In a many-to-many application, many people are able to communicate with many other people. This capability is used today in electronic conferences, where participants "post" notes as though on a public bulletin board and reply to the notes of others. The application helps people look through subjects and keep track of notes and replies.

Combining these multipoint communications capabilities with emerging networking technologies over the next few years will result in new networking applications in the following areas:

computer-integrated telephony, such as online directories, automatic dialing, computerized message desks, intelligent telemarketing, and voice authentication for credit-card operations—offered either through physical integration or voice and data or through logical integration at the applications level

distributed transaction-processing systems

computer-integrated interactive video, for entertainment, shopping, and learning

value-added networks, like today's Dow-Jones and CompuServe information services, such as on-line encyclopedias or electronic news delivery.

### Developing Distributed Applications

In addition to purchasing distributed applications from others, organizations will need to be able to develop their own applications, or to customize others' applications to meet changing business needs.

An increasingly rich set of tools make it easier for the user to develop new networked applications. In addition, a growing body of applications standards will make it possible for applications to work together, and to call one another directly.

OSI has announced a number of standards for applications, including X. 400, for electronic mail applications using public and private networks; FTAM (file transfer and access management), for exchanging files among different systems; VTP (virtual terminal port), which allows the user to access multiple hosts; and JTM (job transfer manipulation), a remote job-entry protocol.

In addition, seven leading computer companies recently formed an international foundation to develop a completely open software environment to make it easier for people to use applications from many different developers. The Open Software Foundation (OSF) is working to develop standard application interfaces, advanced system extensions, an open operating system, and a standard user interface.

Emerging applications standards will allow computers to work together, in routine transactions, without human intervention. For example, distributed computers will be able to work together automatically to check inventories, order supplies, confirm shipments, and make payments. People will still have to negotiate prices and determine reorder levels, but much of the time-consuming paperwork and documentation will happen in an electronic dialogue between computers. It is my personal hope that networked applications within and between companies will finally realize the long-deferred "computer promise" of reducing paper consumption.

New development tools and techniques will increasingly blur the distinctions between end users, developers, and MIS managers. Users will be given more tools to "tailor" applications and interfaces to match their personal preferences, and developers will routinely include "hooks" for integration with other applications and remote network devices for MIS managers to plug into the corporate computing environment.

To eliminate the need for users or applications to have to know where computers, databases, printers, and other network resources are located, some computer vendors are already providing distributed system services that automatically connect users and applications with network resources. For example, distributed naming services allow common logical names to be given to files throughout the network. Distributed file services give a person with the appropriate clearance access to any file on any computer on the network. Distributed queuing services allow any computer to input to any peripheral queue on the network.

Remote network-management services, such as the ability to remotely manage dispersed personal computers (i.e., to update software, to download new applications, to back up files), are available today and will become increasingly sophisticated in order to support tomorrow's new distributed applications.

As open standards lower the cost of entry to the market for networked applications, and as corporate development staffs and major computer vendors intensify their efforts to take advantage of the network, distributed applications and services not yet imagined will put a whole new world at our fingertips.

### The Social Future of Networking

Networking in the 21st century will be quite complex, yet it will appear very simple. It will be as commonplace in our daily lives as electricity, providing as many diverse kinds of applications. It will have tremendous impact on our social and cultural institutions.

We will have to examine and manage the social changes of networking no less than the technical. Networking has the power to allow everyone to participate in a worldwide marketplace—will we be able to ensure that everyone has equal access to it? Networking makes it feasible for people in organizations to share information freely and frequently—will we be able to release ourselves from "chain of command" organizational structures to take advantage of this capability? Networking will give people access to vast libraries of historical and up-to-the-minute written, visual, and oral information—will we be able to develop tools to allow people to chart their own courses of learning and discovery through so much information? Networking has the potential to connect all the world in one global electronic civilization—will we be able to sustain a diversity of cultures?

Today much of the potential of the networking technology that exists is not being realized, because social change has lagged behind. Networking is at once a cause, a result, and a potential solution for many of the social changes we face. I propose that it can also serve as a metaphor to help us create the kind of social flexibility and stability we need in an ever more rapidly changing environment and to think in new ways about the kind of future we can have.

### The Age of Networking

It is a common theme that the economic basis for human society has undergone several major changes. In the past 200 years, the United States has gone from an agricultural society to an industrial society to what some call an information society. I believe we are undergoing yet another major change: we are becoming a networked society.

In an agricultural society, work was essentially manual, directed by nature and (to a lesser extent) by feudal lords or field bosses who simply tried to get as much work out of people as possible.

In industrial society, work was broken down into component tasks that, when assembled with the specialized tasks of other people, produced a product. People were directed by bosses who had attained overall knowledge of the specialty, who in turn were directed by bosses with a slightly broader picture of how things fit together, and so on, until eventually you came to the few masterminds at the top who had analyzed the product and broken it down into component tasks. The efficiencies gained in mass production rose with the volume of goods produced, and the output of goods rose dramatically in the industrial age. However, the role of managers in the industrial age was more than just getting people to produce as much as possible—it was also to make sure they followed directions as precisely as possible. Creativity and flexibility were liabilities on an industrial assembly line.

In the information age, society moved from processing material to processing information. New machines—computers—helped us process ever-larger amounts of data, but work was still organized and performed in an assembly-line fashion. The exhortation was "think" rather than "produce," but the measurements were still productivity (instructions per second) and accuracy (data integrity). Work continued to be broken down into isolated specialties that came together in a meaningful way only at very high levels of the organization. How much you knew, or how much information you had, was directly tied to your position in the organization. As a result, information was passed down only on a "need to know" basis, as determined by the individual's immediate manager above.

In a networked age of rapid, efficient two-way communication, work is still dependent on information *processing*, but there is a new emphasis on information *sharing*. The goal is efficiency in meeting predetermined ends by creating satisfactory outcomes between distributed people or agencies. And the measurement becomes not only the quality of the goods or the information, but also the effectiveness of the work process. The manager's role becomes one of facilitating relationships between people whose shared work adds value to the organization's product.

### From Competition to Cooperation

The language we use to describe the flow of work is becoming more horizontal and less vertically hierarchical. Witness the adoption of Michael Porter's value-added chain, the way "networking" has replaced "moving up the ladder" as a personal career strategy, the use of "reciprocity" as a principle in the emerging economic policy of the European Economic Community, the proliferation of "partnering" relationships, and the ubiquity of the term "trading partner" in place of the increasingly indistinct categories of "supplier," "customer," and "competitor."

The classic one-way economic flow, with a supplier and raw material at one end and a finished good and a customer at the other, is becoming an interwoven, multipolar web of two-way communications, with each entity looking for and creating new opportunities in markets with increasingly indistinguishable boundaries.

Networks accelerate the shift away from production and consumption toward more perfect realization of the ideal of "joint wealth creation." To succeed in the networked future means not just competing, but cooperating.

### Managing the Self-Managing Organization

Just as peer-to-peer networks have proved more efficient than hierarchical ones, organizations that facilitate direct peer-to-peer interaction across functions and across levels of management will prove more successful in the 1990s. To take advantage of the power of the network, decisions will increasingly be made by the people closest to the work. To respond effectively to new markets and opportunities, people will increasingly be defining their own jobs in the act of doing them.

Yet the organizational structures of many companies continue to resemble those of 19th-century industry, or the military chain of command developed for fighting battles with large armies. And despite many discussions about reducing overhead and streamlining organizations, in practice American business actually seems to be increasing the number and the levels of managers. From 1978 to 1985, the entire GNP rose 18 percent, that of blue-collar workers dropped 6 percent, and that of white-collar workers increased 21 percent. In 1985, the real business GNP rose by 2.7 percent, but executives and managers were hired at a rate of 5.6 percent—and administrative staffs increased 3.5 percent (30 percent faster than overall output). Some economists estimate that as much as 40 percent of the "Japanese advantage" is due to lower white-collar overhead.[5]

It is interesting to note that the military enjoys the luxury of operating in a strictly hierarchical manner only during peacetime; in battle, this organizational structure is immediately augmented by one that provides for great latitude in decision-making in the field. Likewise, in many companies today there are actually two organizational structures: a relatively stable hierarchy to do administrative tasks, and short-lived, cross-organizational teams that cooperate to do the work. At times the goals of the formal hierarchy ("making the numbers for our division this quarter") come into direct conflict with the goals of the enterprise ("getting the long-term business"). These conflicts have real consequences for both the individual and the organization. Clearly, common access to information by both interested groups can help define management solutions.

Networks can offer us new models of management as well as help us to implement the management structures we choose. Networks can help managers create and stay plugged into a more self-managing organization by extending their personal network.

In networked companies it becomes easier for people to find one another, to work together regardless of where they are located, to accomplish a task, and then to disband their "virtual" organization when the work is done. And because each person in the network extends the network to other people, access is compounded as physical barriers are transcended.

Networks also make it easier for managers to "know what's going on." Shoshona Zuboff, author of *The Age of the Smart Machine,* notes that networks "make it possible for knowledge to be both centrally located and widely distributed. Computers render moot all the age-old debates about whether to centralize or decentralize, because now you can have all the capacity of extreme centralization and extreme decentralization at the same time."[6]

In a networked organization, managers are able to check on projects and people more directly. They have access to up-to-the-minute information, and they can create and communicate a vision for the future.

The job of the manager becomes one of creating an environment that encourages people to participate in setting organizational goals, to take responsibility for meeting those goals, to establish their own teams, and to determine as a team how best to achieve those goals.

I do not mean to say that a manager who creates a forum for participation no longer needs to make decisions, or that everyone needs to agree before a decision is made. Management must decide frequently what the purpose and the goals of the organization are, and must communicate those decisions.

Open and honest discussion creates the context for people to go forward, understanding the tradeoffs and the intent behind a decision. Early in my career, I noticed that the decisions made out in the open—even those generating dissension—were the decisions that "took."

Peter Drucker calls creating the new organization "the managerial challenge of the future," foreseeing that it "will require greater self-discipline and even greater emphasis on individual responsibility for relationships and for communications."[7] As organizations become more self-managing, I believe they also become more self-rewarding. Participation can be its own motivation, and freedom to set your own direction can be its own reward.

The immediacy and the interconnectedness of the networked economy will both require and make possible a new kind of empowered organization. But networks alone will not create the organizations that can best take advantage of their potential. Managers become more visible, and their personal behavior becomes more crucial in setting an example for the rest of the organization. Today, everyone has a telephone. In some companies, people will take the risk and the responsibility of picking up the phone and trying to fix something they think is wrong; in others, precedent and protocol will prohibit and protect people from having to do anything about a problem.

There is no question that tomorrow everyone will have access to networking technology that will facilitate direct, open, and honest communication across all parts of an enterprise. The question is: What will we do with it?

### The Self-Regulating Global Economy
In a world where goods and services are increasingly represented in digital form, "mass-customized" to meet diverse preferences, and delivered via the network, traditional regulatory

divisions, like national boundaries and market sectors, lose their meaning.

The attributes that once described a nation's economy—what it exports and what it imports—no longer apply. As far as stock exchanges are concerned, it is clear that no material is being transported. Information is the new import and export, traveling at the speed of light across national boundaries, with no tariffs attached.

As multinational entities, most corporations further blur national boundaries—and traditional industry and market-sector distinctions as well. Divisions of General Motors and General Electric have become leading financial companies. Sears, the mail-order pioneer, issues credit cards. American Express, the credit-card company, sells products by mail order. The inserts that come with your credit card bills are like department stores. The department stores issue credit cards and offer banking services similar to personal checking. American Airlines is selling internally developed telecommunications software and training services.

Regulating information as a commodity is paradoxical. Information certainly has value: it can be associated with gain or loss. Yet the value of information is certainly not quantitative. In fact, there is so much "worthless" information available to each of us that intelligent pre-sorting of it, whether performed by people or by machine, is an increasingly desired service. The value of information might be better related to its "timeliness," a qualification no less subjective than "usefulness." What is useful to me may be worthless to you (unless, of course, you determine that it is useful to me and sell it to me).

Information can be dangerous, too. Both organizations and individuals will need to have some kind of access, control, and recourse to correct the information that is collected about them and then circulated on the network. New rules and methods of arbitration will be needed to protect people from the power of the network.

Whereas once the market was a physical place where people had to meet in order to exchange goods, today bartering takes place on electronic communications links, and corporate assets reside in databases. Today's regulations are misaligned with today's 24-hour, globally networked market. New regulations are required, yet it is clear that rapid change will continue to outstrip static regulations imposed from outside. Some kind of ongoing, built-in self-regulation will be necessary. Like the competitors, customers, and users who are cooperating to establish open, international networking standards, it seems likely that the global economy will increasingly be based on participation in markets created by multilateral agreement.

### Relativity and the Age of Networking

Networking, as we have seen, is necessarily a joint endeavor— a group adventure. Previous theories of business management, like earlier theories of physics, were based on the assumption that the universe was a mechanistic place with simple one-way flows of cause and effect. You looked at a problem, analyzed it, determined a course of action and then implemented it. In a networked world, as in the world of quantum physics, things move much faster and "everything interacts with everything" in very subtle ways.

Building large networks of diverse people, processes, and technologies is not a matter of being able to predefine and impose a lasting solution from outside. Networks must be planned and managed from the inside out. As Peter Huber says, "The old network had a simple Euclidean structure, with an inside and an outside, and clear divisions between them. The new network is described by the mathematics of fractals, with nodes leading into lines, which lead into more nodes, the pattern replicating itself indefinitely down to the smallest scales."[8]

Networking is by its very nature an ongoing and self-perpetuating process. As Huber says, ". . . networking is contagious. Each new connection . . . creates opportunities for two more."[9] To take advantage of the potential of networking means to have the ability to start small—with anything, anytime, anywhere— and connect to other networks in other places in a smooth modular fashion. To support this need to connect and integrate as you go, networking technology must become ever more "scalable"—that is, congruent as you go up and down in scale.

The relativity and the interdependency of the networked age mean that organizations can no longer be successful with managers who stand behind people and direct; they need managers who will get out in front and lead. In the industrial age, managers were able to base their decisions on "what worked in the past." More recently, managers have based their decisions on what works in the present—"what happens to the bottom line." In the networked age, managers will base their decisions on the future they want to create. Their job will be to set the direction and maintain the flexibility to incorporate what is learned along the way.

Someone recently asked me what advice I would give to a CEO trying to understand and utilize the potential of networking in today's business environment. I said: "Think about what you need, begin to build it, and add to it over time. Don't try to do it perfectly. Just begin to do it. Don't try to implement the grand, global scheme all at once. You can't. By the time you get there, it will have changed." I realized later that this was precisely the opposite of the good advice I had been given as a new manager, fifteen or twenty years earlier. When did the change happen? I can't say. The future, after all, is not "something that happens to us," it is the collective results of the choices each of us makes every day.

### References

1. Tom Valovic, "Private and Public Networks: Who Will Manage and Control Them?" *Telecommunications,* February 1988.

2. "Network Communications Support Service: Market Trends," The Ledgeway Group, 1986.

3. "The Cost of Network Ownership," The Index Group, Inc., 1987.

4. Peter W. Huber, "The Geodesic Network: 1987 Report on Competition in the Telephone Industry," January 1987.

5. Lester D. Thurow, "A Positive-Sum Strategy for Productivity That, Thurow Says, Does Not Add Up," *Scientific American,* September 1986.

6. Gary Emmons, "Smart Machines and Learning People," *Harvard Magazine,* November-December 1988.

7. Peter F. Drucker, "The Coming of the New Organization," *Harvard Business Review,* January-February 1988.

8. Huber, "The Geodesic Network: 1987 Report on Competition in the Telephone Industry."

9. Ibid.

# 6

# A Multi-Dimensional Look at the Future of On-Line Technology

*Al McBride and Scott Brown*

There will be increasing competition during the 1990s because of aggressive global competitors and a trend toward lower trade barriers. As companies struggle to differentiate themselves, time will become an increasingly important strategic asset or liability. It will be imperative to design, build, sell, and service products faster, and to do it at a lower cost.

Companies will be able to simultaneously reduce costs and improve services by using on-line technology, which enables a user at a terminal, a personal computer, or a workstation to immediately access and update information in a database, at any time, from any location. This technology is a tool that will allow organizations to respond to competitive pressures.

An example of the use of on-line technology is provided by the California Department of Motor Vehicles (DMV). The DMV may be the largest information provider in the state of California by the year 2001, providing information to state and local law-enforcement agencies, insurance companies, financial institutions, and the court system. The DMV envisions a paperless system in which vehicle and license renewal, payment of fees, and traffic citations will be handled electronically. Digitized color photos and fingerprints will be added to the database to speed the renewal of drivers' licenses. In this way, the DMV can cut down the number of people who must drive to a DMV office to conduct business by making it possible to conduct routine transactions over the phone.

The California DMV's use of technology to improve productivity and reduce costs is an example of the trend toward moving information on line, which began in the 1980s and which will continue to be a dominant theme for users of information technology (IT) in the 1990s and beyond. This trend cuts across all industries and all geographic regions. It is driven by the falling cost of technology and by increasing global competition in all industries.

Faced with the need to improve services, reduce costs, and respond rapidly to change, the DMV is upgrading its 20-year-old data-processing system. The database currently holds information on approximately 20 million drivers and more than 25 million vehicles, and is expected to grow by 30 percent over the next 20 years. Moreover, the DMV is constantly being asked to provide new services and to make changes to existing services.

Under the existing system, the changes generated during a single session of the California legislature require approximately 50,000 hours (26 person-years) of programming and analysis. Some changes, such as correlating the driver database with the vehicle database, are not possible without the expenditure of millions of dollars to restructure the database.

The DMV's new system will not only provide 24-hour access to the vast amount of information controlled by the department, it will also give the DMV the ability to respond to change rapidly at a low cost. This new system will use a relational database consisting of tables that can be easily modified to accommodate change. Through the use of high-performance, relational database technology, the DMV will be able to rapidly introduce new programs and improvements to existing programs.

Imagine a California Highway Patrol officer in the late 1990s on a routine patrol of Interstate 280 between San Francisco and San Jose. As the officer drives down the highway, an electronic eye in the front of his car reads the plate numbers of the cars on the road. This information is automatically transmitted to a DMV data center, where information about each vehicle is retrieved and analyzed. If the system identifies a vehicle that has been stolen or involved in a crime, that information is relayed to a computer screen on the dashboard of the officer's car. If the information indicates that there may be dangerous criminals in the vehicle, the system also displays the location of other officers on patrol in the area. Those officers are alerted to the potentially dangerous situation and begin moving toward the officer needing assistance. The exact location of each officer is tracked via satellite and displayed on each officer's dashboard screen. Such an application of on-line technology would allow Highway Patrol officers to provide better protection to themselves and to the community.

The DMV is testing an on-line system that will allow people to make appointments for driving tests and vehicle registration renewal by using an 800 number. It is also evaluating the use of credit and debit cards for the payment of fees, including parking fines and fines for moving violations that do not require a court appearance.

In the business world, companies are responding to increasing competition by enhancing productivity and reducing costs through the implementation of fast-cycle capability. Time becomes a source of competitive advantage, and time-based competition requires on-line access to information throughout an organization, from order entry, to purchasing, to engineering, to production, to shipping, to service.[1] Eventually, the entire enterprise will be on line. These enterprise-wide on-line networks will consist of a variety of equipment from different vendors. Geographically distributed operations, suppliers, and customers will all have access to up-to-the-minute information.

The 1990s will become the decade of the on-line enterprise. In the 21st century, on-line technology will be as commonplace as the telephone and the personal computer are today.

In the retail industry, on-line reordering and stocking will become the norm. In manufacturing, suppliers will be connected as part of the on-line system; engineering will be tied in so that changes can be made in the factory within minutes. The securities industry has already begun to offer 24-hour stock trading. Public networks in the telecommunications industry will offer users "instant service," allowing new or changed services in minutes, and a "self-healing network," with on-line, automatic fault isolation and call rerouting. In the gaming industry, future casinos will be completely electronic, and plastic cards will replace cash and chips.

New and flatter management structures become possible as more information within an organization comes on line.[2] Organizations will no longer be forced to choose between centralization, for tighter control, and decentralization, for faster decision making.[3] On-line technology will make it possible to have centralized control with decentralized decision making. Databases in the on-line enterprise will hold large amounts of complex data, such as digitized voice, image, graphics, and full-motion video. This will lead to a very rich set of transaction possibilities. For example, it will be possible to send an electronic memo including a video "clip" of next month's ad campaign.

## The Requirements of On-Line Processing

On-line processing requires instantaneous response to requests that occur at unpredictable times and rates. (For example, economic news in the morning paper may generate unusual stock trading as investors rush to modify their portfolios.) On-line processing requires the following:

• Continuous access by many users to large amounts of distributed data. The underlying technology must provide high system availability, the ability to repair the system while it is on line, high performance to handle unpredictable peaks in utilization, large storage capacity to accommodate the massive corporate database, and high-speed communication for rapid response.

• The ability to accommodate rapid and unpredictable growth in the size and use of the application. The underlying technology must support system expansion without lengthy software conversions. Many on-line applications generate high growth and require the ability to increase capacity without significant system outages.

• The ability to manage large networks of heterogeneous computers. Most companies have a heterogeneous computer environment, with many different types of equipment purchased from multiple vendors. One of the biggest challenges of the 1990s will be the management of this resource.

• A good environment for developing strategic applications. Most companies realize that in order for information technology to provide a competitive advantage, they must develop and maintain their own strategic applications. Using application tools to develop these strategic applications speeds development and maintenance and ensures high quality.

## The Multiple Dimensions of On-Line Technology

In the 1990s, enterprises will move toward more on-line data processing to support decision making and operations. The hardware and software components of their computer systems, referred to collectively as the *technology*, will continue to ad-

vance and be shaped in ways consistent with demands of the
on-line enterprise.

The challenge for the 1990s will be to correct the basic imbalance of the computer industry: technology is well ahead of
operating systems, which are ahead of applications, which are
ahead of users. These multiple dimensions (called *domains*) of
on-line technology, and their current and future emphases, are
shown in figure 1.

Balancing these four domains will allow new functionality in
one area to be matched by appropriate advances in the other
areas in order to provide cost-effective and pervasive benefits.
Often, advances in a specific field are based on ideas conceived
years ago, but whose time has come as a result of other advances. RISC (reduced instruction set computer) technology is
an example. The design principles to be used in RISC were
under investigation in the mid-1970s. Thus, as we look forward, we should also look back. The history of the computer
industry is full of examples of conceptual breakthroughs that
become products years later.

Each of the four domains embraces an independent trend.
Computer technology presses for higher performance, greater

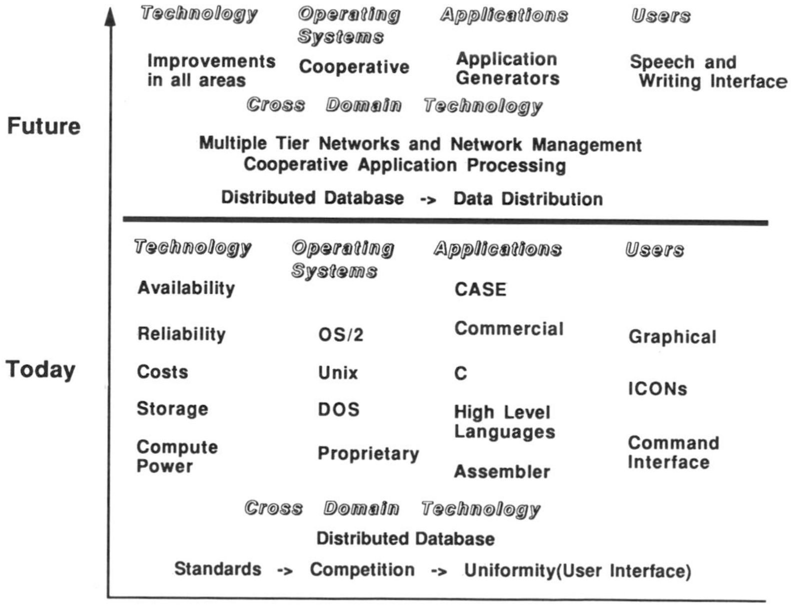

**Figure 1**
Multiple dimensions of on-line technology.

reliability, and lower cost. Operating systems aspire to standards and cooperative interactions between multiple systems. The applications area pursues ease of development and portability. Finally, graphical user interfaces strive to improve users' ability to interact with the systems in an efficient, easy-to-use, and memorable manner. The future holds continued improvements for each trend within each domain.

Cross-domain developments will serve to strengthen the overall functionality of new systems. These developments include distributed computing based on robust data transport and sharing across heterogeneous networks. The future of the on-line enterprise lies in a distributed environment. In this environment, computing resources are located where the work is being done, and data moves to and from these resources in a controlled, user-transparent manner.

### Trends in On-Line Technology

There are five major trends that address the requirements of on-line processing. The first is growth in microprocessor performance. Is there an end in sight? If not, what is the next hurdle, and how might it be overcome? Microprocessor performance advances generally signal forthcoming developments in other areas. Processor performance is a major determinant of a system's ability to provide a fast response to a request or an adequate response to a very complex request. The second trend consists of the new requirements for continuously available systems. The third trend is the cross-domain endeavor of distributed database technology. The fourth is the evolution toward three-tier distributed on-line enterprise networks. The fifth addresses the user's need to control these technologies in a network, using network and systems management. All these trends directly address the requirements of on-line processing and will determine the pace at which on-line processing will progress.

### Trends in Microprocessor Performance
In the 1990s we will witness astonishing microprocessor performance improvements. These improvements will further reduce the cost of computing, lead to new image-based applications, improve the response times of on-line systems to users' requests, enable a system to serve more users, and allow people to interact with computers in more natural ways (such as

through voice and handwriting). These powerful microprocessors will be used to improve other parts of our computer systems, and thus to make possible new applications. For instance, the data to be transferred to disk or to another computer will be compressed by microprocessor and decompressed upon use. The cost in time of the compression/decompression sequences will be more than saved in the reduced transmission time of the compressed object. Further savings will be realized by the reduced data-storage requirement of the compressed objects.

Four questions come to mind regarding microprocessor performance: What is achievable? What factors may constrain future performance improvements? What will succeed RISC? How will current systems migrate to new processor technologies?

### *What Is Achievable?*

Since the 1970s a series of new microprocessor architectures have arrived, with improved performance and with designs matched to the technological advances of the day. In the early 1970s came the 1–8-bit CISC (complex instruction set computer) microprocessors. Next, 16- and 32-bit CISC microprocessors followed in the late '70s and the early '80s. RISC (reduced instruction set computer) was introduced in the mid-1980s. The mid-1990s will see yet another architecture, which we will call the New Instruction Processor (NIP).

The first 1–8-bit microprocessors, developed at IBM and Intel in the early 1970s, used the technology that the first pocket-sized calculators and digital watches used (*n*-channel metal oxide semiconductor, or NMOS). The small data width was necessary so that the design could be implemented on a single silicon chip and packaged in the ceramic carriers of the day.

In the late 1970s and the early 1980s, CISC microprocessors, such as Intel's $80 \times 86$ and Motorola's $680 \times 0$ families, were introduced. These microprocessors are the core element of virtually every personal computer and workstation made since the late 1970s.

In the 1970s, the high cost and limited size of memory made efficient use of it a prime system-design requirement. Reducing the memory required to store a computer program was a prime objective of CISC processors. To accomplish this, numerous complex instructions (such as "move *n* bytes of data from address *x* to address *y*") were defined in the architecture. During program execution these instructions were fetched from memory by the microprocessor and executed through a series of simpler steps under the control of another program (called *microcode*) contained on the microprocessor chip. This design philosophy maximized the utility of limited memory bandwidth while keeping the memory and system costs down. The design tradeoffs reflected in these microcoded processors limit today's CISC processors from making significant performance leaps as new technologies emerge.

During the mid-1980s, the first commercial RISC microprocessors were introduced by Mips Computer Systems, Sun Microsystems, and Fairchild Semiconductor. RISC improved on CISC by eliminating the microprocessor's on-chip microcode program and by implementing fewer and simpler instructions. RISC thus achieves faster instruction execution at the expense of more instructions and memory consumption per program. Given the reduction in memory costs over the last decade, this tradeoff was appropriate. At the expense of doubling the program memory, a threefold improvement in performance is achieved.

RISC was architected and designed to be implemented in the new technology of the 1980s: CMOS (complementary metal oxide semiconductor). RISC microprocessors achieve high instruction execution rates by simplifying the instruction decoding. All instructions are designed to fit into a few fixed-length 32-bit instruction formats, which execute in one processor cycle in most cases. RISC achieves approximately a threefold performance advantage over CISC when implemented in the same technology. As technology advances yield smaller and faster devices, RISC—with its reduced design complexity—is better able to take advantage of the technology to improve its performance.

A new processor architecture will arrive in the mid-1990s that combines the best of CISC (short average instruction length) and RISC (one-cycle instruction execution at high speeds) architectures to yield a design that is optimized for performance above 50 megahertz. At very high processor speeds, memory costs and bandwidth again become a primary design consideration, as they were with CISC. A two- or three-level memory hierarchy (versus CISC's one level) is required to deliver instructions to the processor to keep it busy. This NIP (new instruction processor) architecture will probably feature a simpler design implementation for ultimate instruction execution speed and a reduced average instruction length. NIP will minimize the cost and maximize the utility or bandwidth of the two new layers in the memory hierarchy.

The simpler design will allow NIP to be implemented in the high-frequency technology of the day—probably bipolar emitter coupled logic or gallium arsenide, both of which have reduced functional density when compared to CMOS. The reduced average instruction length better optimizes the memory design and processor speeds. This will result in microprocessor speeds in excess of 100 MIPS in the early 1990s, scalable to 400 MIPS by the mid '90s and 1 BIP (1000 MIPS) by the turn of the century. These speeds will be achievable two years earlier and at half the cost of comparable RISC designs.

Since the early 1970s, each generation of microprocessors has brought greater performance and ushered in a wave of new products. It is expected that CISC will evolve at a 26 percent compounded performance rate improvement, while RISC will achieve up to a 50 percent improvement in MIPS per year. However, it is our opinion that, above the 100-MIPS range and at operating frequencies in excess of 50 MHz, a more efficient architecture will be required.

By the year 2001, technology will be capable of producing a 1-BIPS (billion instructions per second) microprocessor, whereas less than 20 years earlier the state of the art was 1 MIPS. This will be a thousandfold improvement. It should be noted that while a 1-BIPS processor is achievable, CISC will achieve 50–100 MIPS and RISC 200–500 MIPS by the end of the century. CISC will continue to be the lowest-cost microprocessor design, with RISC and NIP more expensive because of the elaborate memory hierarchy needed to sustain their performance. We must know how to use this power, or there will still be an imbalance of technology in the year 2000.

### What Factors May Constrain Future Performance Improvements?

RISC achieves higher performance than CISC by utilizing a longer average instruction length and requiring a larger program memory. RISC's biggest limitation is its fixed 32-bit instruction length. This results in program lengths of more than double what is otherwise required.

Longer programs increase system cost and limit system performance. System cost is increased because more storage space is required for programs.

System performance is affected as follows. As the speed of the processor increases, the memory system needed to sustain the processor's performance will become more sophisticated with the introduction of one or two cache levels. *Caching* is a name for the techniques and algorithms that best utilize the additional levels of memory. Each level of memory closer to the processor is faster, more expensive, and smaller than the level above it. Each level is designed so that the data or instructions the processor needs are likely to reside in it. The percentage of the accesses that are satisfied by a level is called the *cache hit ratio,* and its opposite the *cache miss rate.* If the desired data do not reside at a certain level, the slower level above it will have to be accessed to obtain the data. During this time the processor is idle, waiting for the data to be retrieved.

The extra memory requirement of longer programs is felt at each level of the memory hierarchy, but especially at the one or two new cache memory levels between the main memory and the processor. Expensive, high-speed cache RAM (random access memory) components are needed to store enough long instructions to maintain the 90%+ cache hit rate needed to sustain processor performance. Ultimately, longer programs increase cache misses, causing a decrease in performance.

Figure 2 plots the processor performance (MIPS) versus the cache memory operating frequency for a two-level memory system (main memory and one cache level) to illustrate the performance degradation that occurs. The degradation is due to cache misses and limited memory bandwidth.[4] This simple model shows that at high frequencies of operation, where the cache memory limits the sustainable microprocessor performance, shorter instructions yield almost a twofold performance improvement. Furthermore, shorter instructions will also allow the processor and its memory to be implemented on the same silicon chip versus separate chips earlier, a necessary step to achieve super operating speeds above 200 MHz.

### What Will Succeed RISC?
In the early 1990s, RISC will mature and enter the multiple-instruction-execution phase reached by all processor architec-

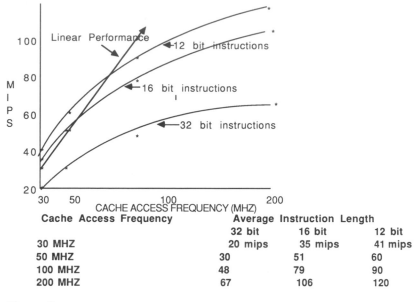

| Cache Access Frequency | Average Instruction Length | | |
|---|---|---|---|
| | 32 bit | 16 bit | 12 bit |
| 30 MHZ | 20 mips | 35 mips | 41 mips |
| 50 MHZ | 30 | 51 | 60 |
| 100 MHZ | 48 | 79 | 90 |
| 200 MHZ | 67 | 106 | 120 |

**Figure 2**
Achievable performance in complex multi-tasking environment.

tures. The NIP design will take advantage of a new high-performance technology that will be optimized for high-frequency operation but will have relatively low density. Such a technology is bipolar *emitter-coupled logic* (ECL). This is the technology of all high-performance mainframes today. Recent advances in semiconductor processing have enabled a complete microprocessor to be implemented on a single ECL chip for the first time. The features and limitations of this technology will dictate changes in the microprocessor architecture.

ECL circuits are superior to CMOS in two ways: ECL circuits operate 3 to 6 times faster than CMOS, with less speed degradation due to operating temperature and voltage variations. Furthermore, ECL chips can transmit signals to memory chips much faster than CMOS. However, ECL chips, due to power dissipation and the number of devices per circuit, can achieve only about one-third the functional density of CMOS.

### How Will Current Systems Migrate to New CPU Technologies?
One of the questions associated with technology advances is: How will the current applications move to the next generation? Will applications have to be rewritten? Vendors of software and

operating systems must solve this problem to avoid becoming victims of their past success.

Because of the cost/performance improvements of RISC, vendors must make the transition of new platforms. User-transparent migration is not an insurmountable problem, and the vendors that solve it will determine the pace at which the 1-BIPS microprocessors are incorporated into systems. Progressive vendors will provide for their customers' software moving to the new processors so that both old and new applications can profit from the improved microprocessors.

Numerous approaches can be taken to ease migration. For example, emulation of the old architecture by the newer, faster architecture has been done in the past. This allows old applications to execute at improved performance rates, but not to fully realize the benefits of the new technology. Recompilation of applications to the new system is another, although not an optimum solution. A developer may not have the source code. Or if the application contains assembler language sections, this approach could involve a partial rewrite. At the very least, a new software quality-assurance cycle will be required before deployment.

An interesting possibility is an attached-processor approach, where an old and new processor share memory, and application execution is shared. Applications and the operating system both execute on the processor compatible with their program files. Over time, more and more of the code is converted (recompiled) to the faster processor.

Standards such as the C programming language and the Structured Query Language (SQL) for relational databases offer improved migration scenarios for users and vendors. They allow applications to be written using an industry-standard programming language and relational database file command language. These applications then can be recompiled and executed on a new microprocessor platform with few or no changes. However, many vendors enhance the standard C and SQL implementations with features specific to their hardware or operating systems. Compatibility within a single vendor's offerings is probably ensured, but migration to another vendor's software or hardware may be difficult or even impossible. Adherence to open standards can mitigate this.

At least four software porting approaches are possible. Expect two architecture changes before the year 2001: first to RISC and then to NIP.

## New Requirements For Continuously Available Systems

System availability will be more closely scrutinized by enterprises as they perform more on-line processing essential to business survival. Availability is but one of three attributes that all future systems will possess; the other two are reliability and expandability. *Reliability* is a measure of the likelihood that a component in the computer will fail and/or require service. *Expandability* means that the system can grow easily, transparently, and cost-effectively to handle a greater workload over time. *Availability* is a measure of the degree to which an user experiences uninterrupted service from his computing system.

With 100 percent availability,[5] after a component or software failure the system continues to operate, and an on-line service call can be completed before another failure occurs. Availability requires fault tolerance, which is the system's ability to operate in spite of component failures and service that is both timely and performed without having to interrupt use of the system. Availability will take on new meaning as a result of distributed computer networks and the application models they will make possible.

Five unique application environments will coexist, and each will require a different high-availability solution that is automatically invoked when the application is initiated. Today, vendors offering high-availability systems and software address but one of these five environments. That one is the host (mainframe or minicomputer) resident application and data communicating to a directly connected terminal. The terminal function may be performed by a personal computer emulating a specific type of terminal.

These five application models will be deployed across three-tier computer hierarchies. The three tiers will include the central database or host, the local-area network (LAN) server, and the workstation or personal computer. The first tier will include multiple hosts with wide-area network (WAN) connections between them.

WAN long-distance connections have a limited data-transport bandwidth. The transmission rates vary from 9600 to 2 million bits per second, depending on the amount of money one wishes to spend for communication line usage. LAN connections are thousands of meters in length and can transmit data at 4 to 10 million bits per second. More expensive fiber optic LANs can transmit at 100 million bits per second. Both of these bit rates are far less than the data-transmission rates within the average personal computer. For example, to transmit a 10 × 7-inch black-and-white picture from one computer to another requires the movement of approximately 600,000 bits of information. Compression techniques can reduce the number of bits, but communication protocols add to it. Moving the image will take approximately 60 seconds over a 9600-baud line, 10 seconds over a 64,000 line, or ¼ second over a LAN. A four-color picture would double the times. If multiple people use the same line, queuing delays can be more significant than the transmission times. Taking into account or attempting to hide these relatively slow transmission rates from applications and their users is a challenge to both distributed database technology and application developers. A poorly designed application across two or more systems connected via communication lines will result in poor user transaction response times due to communication delays.

The first application environment, the terminal-to-host model, is the oldest. In this model, the application and the data reside in the host and communicate with a remote terminal for input and output. An example of this is an automatic teller machine (ATM) application, where the customer uses a card and the terminal's keys to instruct a remote computer. The ATM communicates with the remote computer over telephone lines. The computer validates the request against the account database, writes messages to the ATM, and triggers the dispensing of money. In a distributed environment, the ATM will communicate to the remote host via a branch LAN server that also does local processing for the branch. To gain high availability, it will require fault-tolerant communications support in the server and the host.

This computing model will continue to be prevalent for three reasons. First, to change existing applications written for this environment, code must be rewritten, and this may not make economic sense. Second, until distributed database technology advances (covered in the next section), it will be risky to allow strategic data onto the network. Finally, applications that interact heavily with a large database perform better when the application and the data reside on the same system. Transaction response times are better, even given a relatively slow communication path to the user's workstation.

The second model is that of the stand-alone workstation or PC. This is the normal PC operating mode when we run our word-processing or spreadsheet software. Here, the prime concern is the protection of data originating in the workstation. This problem is best resolved by connecting the PC to a local LAN server or to a host via a direct connection. The server or host provides a data backup or data mirroring service, so that if there is a disk failure at the workstation or if the user inadvertently deletes a file, the files can be retrieved from the server or host.

The third application model is that of the file server. Here the application runs in the workstation, and the server provides file services. In this model, the file server should be fault-tolerant or an additional hot standby server should be present and attached to the LAN. In the latter case, when the active servers fails, the standby server should take over in a manner transparent to the application. This model is used when data must be shared among the users on the LAN, as with electronic mail.

The fourth and newest model is cooperative processing. Cooperative applications aim to divide the application processing between the workstation and the server. The division attempts to minimize the data transfer between the portions of the application to improve user response times and to allow more users to communicate over the LAN. This model has come about out of the computer industry's attempt to improve the user interface for applications. A responsive graphical interface requires the processing power of the workstation, but the data to be processed must be shared and protected, and/or cannot physically reside on the workstation. In this model, the servers, in order to be fault-tolerant, must be able to continue the execution of the server-based portion of the application in the event of a single server failure. An example of a cooperative application is an SQL back-end database (host or server resident) and a workstation front-end query and report generator.

In the future, a fifth model will provide three-tier (host to server to workstation) cooperative processing. New data services, discussed in the next section, will fit this model, as will new applications. In this configuration, data and portions of the application will exist at the appropriate tier, based on the performance and security required of each portion. An example of this is a branch bank application. When the customer arrives at the bank, the teller enters the customer's identification code into a terminal, where the customer's signature and picture are then displayed. Other relevant information will also be available, such as the customer's birthday or the fact that he just deposited a large sum of money. Most likely the customer account data is maintained centrally on the first-tier computers. The customer's signature, picture, and history with the branch are stored on the branch server, because to retrieve them over a communications line would take too long. If the customer does not normally frequent this branch, the data must be retrieved from another branch. The application, knowing that the data do not reside locally, may opt for only retrieving the signature before continuing customer service to avoid the customer's perceiving the service as unresponsive. The picture is retrieved as a background task during the course of the visit so as to reside locally if needed.

Here, fault tolerance will be provided by a fault-tolerant host and synchronized server execution. The network connections between the hosts will also need to be fault tolerant. Multiple network links to the servers will need to be supported via combinations of leased and dial-up lines, with automatic switchover in the event of loss of service on an individual line. Generally, continuous availability is not required at the workstation level. The rationale behind this is that only one user is affected, who can easily use another workstation to resume work. The workstation will require server data backup, or at least a portable hard disk.

Availability across the three-tier computer network will be accomplished without requiring the application to be written to accommodate it. Rather, the distribution parameters of the application will mark the kind of availability model under which it must execute. In the future, the same levels of availability currently obtainable from a few computer vendors will be available for commodity-based systems. Thus, an entire enterprise network will be fault tolerant, from the workstation to the server and to the host.

### Distributed Database Technology
As computer networks become more pervasive, far-flung, and complex, the need exists for tools and services to tie together

the four domains we mentioned above. Distributed database tools and systemwide directory services are examples. These tools will play a major role in the systems approaching year 2001.

When each of us carries a portable computer, we will insist upon being able to upload and download files to and from it and a remote data vault by simply plugging it into a phone receptacle. We will expect the accompanying systems networking and data management to be transparent.

Future cooperative applications will be split between two or three platforms and will require large amounts of local, current data. New distribution, synchronization, and control problems will need to be solved. Applications that use graphics or images need local data; otherwise the time required to obtain the data over a communication line will prevent their use. Cooperative applications will require that a business know precisely what software is running on each workstation before it is allowed to interact with a central database, in order to maintain security and data integrity.

All the availability and reliability mechanisms designed into three-tier networks will be of little use if the wrong version of software is used to update the on-line database. Today, the software and data residing on personal computers are largely uncontrolled, and only the user knows what he has. Updates are mailed to users in the form of a floppy disk, if at all. Files are shared either by manual transfer or by a series of forgettable steps.

Distributed database technology can remedy this. Today, "distributed" means that one database is divided into local partitions. When the user accesses data located in a partition that is physically close to him, his response time is better than if he accessed remote data. Updates to the database via the partitions are made possible through locking mechanisms that are enforced across the network to prevent simultaneous changes to the database by multiple users. High performance and transaction throughput are best achieved using parallel computing architectures. Parallel computing architecture provides for common access to the database by concurrent transactions that each may be accomplished by multiple processors. The processors involved in a transaction are dynamically assigned to best balance the load on the system and to achieve a linear transaction throughput increase as additional processors are added.

"Atomic" transactions are supported across the network to ensure data integrity in spite of component failure or user error. To be atomic, transactions must be completed in full, or the databases are returned to their state prior to the beginning of the transaction. In the future, distributed database technology will be extended to embrace the data distribution and replicated data issues of the distributed on-line enterprise network.

In the 1990s, "distributed" will include data distribution—the management of copies of data across a network that also allows for the updating of the information, as with the partitioned distributed database. This idea was studied in depth in the 1980s, but it will be implemented in the mid-1990s with the proliferation of cooperative computing resources across a network.

For data to be distributed, new attributes are required. In addition to the actual bytes that represent the information in the database, distributed data will require policies, event triggers, and use restrictions. Policies will include when to distribute and retrieve the data. Synchronous, asynchronous, and periodic delivery will be required, along with instructions as to where in the network tiers the data should reside. Use restrictions will describe by whom and how the application is to be run. Event triggers will describe under what circumstances actions must be taken to prevent failure, or what conditions warrant a message to the higher-level tiers.

The actual data will need to be marked with the compression algorithm and other techniques used to best facilitate data transmission between the tiers. Efficient delivery is required to ensure that the locality and the use of data are the same in a data network consisting of a central data vault with coherent, shared, and distributed whole or partial copies.

A systemwide directory service will free the users and the applications from having to know where the data needed to access it reside. Today, to access data, a user has to know the equivalent of a zip code for each item.

### Three-Tier Distributed Computing Networks

Currently there are three prevalent computing environments: host-based, PC-based, and server/PC based (figure 3). Beginning in the early 1990s, computing environments will be linked much more closely, with each computing component assisting

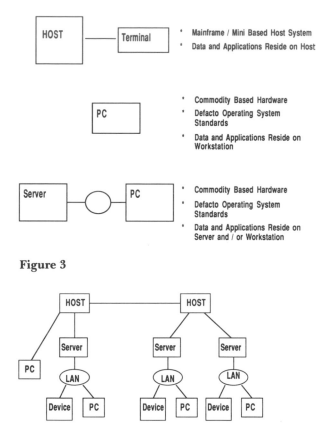

**Figure 3**

**Figure 4**

others to make operations easier and systems more reliable and responsive. The computing environment of 1995 will look like figure 4. The one- and two-tier environments of today will be replaced by an assemblage of linked two- and three-tier models. Within this distributed hierarchical environment, applications and data will reside in locations that will optimize performance, security, and application development. The management and control will be centrally defined but locally administered. In fact, the pace of movement to this distributed environment will be determined by the enterprise's ability to manage the risk, because control is required when moving critical on-line applications and data to local environments. The task of systems management will be greatly lessened if the equipment at each tier (especially tiers 1 and 2) is homogeneous.

Imagine how a branch bank loan application would work in this environment. The customer comes into the local branch and the teller goes through the greeting process of using the customer's identity card to bring up his picture, history, and signature. Next the customer is sent to the loan officer, who signs onto the system and thereby gains special privileges not given the teller. The customer information is immediately displayed, but now it includes his credit history. The loan to be applied for is determined and the terms negotiated. For various interest rates and payment schedules a display (graph) of the amount of money involved is generated by the local server. The rates are obtained from the central database, and the program used to generate the display is verified by the central computer. If a special situation exists, the branch manager signs onto the system with yet more privileges, and he conducts the negotiations using his rates and payment options. All of the data, applications, and bank participants involved in this complex transaction have been centrally registered and given access IDs. The processing and security enforcement is distributed throughout the network so as to not overload the central computers and so as to provide fast response to the customer.

Advances in database technology will be needed to enable enterprises to profit from image- and graphics-based applications that need the performance advantages provided by local servers and workstations. Also, as applications and data move into the network, the reliability and expandability of the servers will determine how applications and systems are designed. Without a reliable server, critical data would need to be maintained at the host, with the incumbent loss of performance when it is accessed over a WAN. The key to acceptance for these new tools will be to enhance central and workstation processing while enabling future local and cooperative processing.

Another possible scenario is that two-tier computing will persist, but that multiple hosts and servers will occupy the first tier and connect with the second-tier workstations over a LAN/WAN network. In comparison with today, this has the advantages and disadvantages illustrated by figure 5. The major drawback is the need for technology which will administer a network of heterogeneous systems. As the number of servers increases, management complexity grows fast, and an enterprise's trust of the system is reduced.

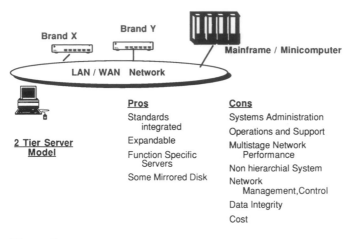

**Figure 5**

Introducing one more tier brings an element of control and symmetry to the network. Future on-line enterprise networks will require the role of host (first-tier) computers to grow and provide more control functions. These computers will definitely need to have the highest availability possible, and they will have to be fault tolerant because of their key role in the network. With the control apparatus in place, enterprises will be able to distribute and enhance their critical business applications.

Application development will shift to the workstation/server platforms. These standards and commodity-based environments promise faster application development and provides the user with a friendly and easy-to-remember interface. As technology evolves, these applications will migrate more easily to a new platform. The role of the host will expand into a central data vault and control center, in addition to that of a network gateway, application, and database platform.

## Network and Systems Management

As the trend toward the on-line enterprise continues, entire organizations will be connected by thousands of terminals, PCs, point-of-sale devices, servers, minicomputers, and mainframes in large, heterogeneous networks. These networks are likely to include both new and old equipment, built by different manufacturers and having differing capabilities. The management and control of these systems will be a major challenge.

This difficult task will be beyond human capability without additional tools and improvements in technology. According to a study conducted by Jim Gray of Tandem Computers, a significant portion of system failures are caused by operational errors.[6] This underscores the importance of network-management tools to assist human operators. Without effective management, computer systems and networks will not reach their potential of bringing entire organizations on line. Other technologies have also faced the issue of management. Telephony and transportation, for example, are far ahead of computer networks in solving their network management problems.

Network and systems management includes the configuration, control, and diagnosis of computer systems, networks, and their environment. The difficulty of the task has increased enormously with advent of the PC. Each PC must be viewed as a miniature data center, because it has its own storage, processing, and communications capability.

Most companies want to control the distribution of software and data files to the PCs in their network. When there are tens of thousands of these devices in a network, this becomes a huge task. The management system must keep track of the current state of each device in the network and allow information about dozens of devices to be simultaneously accessed and updated. In this sense, network management is really an on-line application, requiring high-performance relational database technology.

Currently, network and systems management technology consists of little more than instrumentation. Instrumentation in this case means that it is possible to programmatically configure and control systems and network devices, and that these devices are capable of sending unsolicited messages called *events* to a central collection point. These events carry information about changes in the state of the network, such as notification that a modem has failed or an ATM is out of cash. This basic capability is required before higher-level functions can be implemented.

Instrumentation makes it possible for programs to detect failures, but it is still difficult for programs to determine the real cause of the problem. For example, a program may detect transmission failures but cannot tell whether the communications line failed or whether heavy traffic was overloading the network. Without knowing the cause of a problem, human operators cannot take remedial action. Current management systems are fairly primitive in that they require highly trained human experts to operate them. As networks increase in size and complexity, even these experts will not be able to manage networks effectively.

Network management technology can be expected to evolve from instrumentation to the use of expert systems, to automation, then to simulation. As management systems evolve from instrumentation to expert systems, they will be able to diagnose and correct low-level problems. An *expert system* is an artificial intelligence application that uses a knowledge base of human expertise to aid in solving problems. The degree of problem solving is based on the quality of data and rules obtained from a human expert. This will free human operators to deal with more complicated, higher-level problems. As expert systems improve, management systems will be capable of handling increasingly higher-level problems. Eventually management systems will reach an automation stage where the systems can respond to all but the most catastrophic problems, which will still require human intervention.

Once automation is achieved, it will be possible to build management simulators similar to the flight simulators used to train pilots today. These simulators will obviously be used to train new operators, but they will also have other important uses. The simulators will make it easier to implement new applications by providing a training facility for operations personnel. The simulator will also be used to forecast the cost of making changes to the network. It is possible to envision a time when the costs associated with integrating another company's network will be factored into the cost of a merger. Management simulators will allow those costs to be accurately estimated.

In a heterogeneous environment, it is likely that the operations staff will be geographically distributed and may not be trained on all types of systems in the network. In this case, the management system will be responsible for command-mapping and local language support.

As management systems become more sophisticated, technology will enable an enterprise to control the use of network resources to the point that it will be possible to sell excess computing capacity without fear of compromising security. In this way, companies can generate revenue with idle computing resources in the same way that banks generate revenue by depositing idle funds in overnight accounts.

On the basis of the size and the complexity of management applications, it is reasonable to assume that the management control center will require a very large, high-speed system with lots of disk storage. Management is likely to be the largest and most important application on the network. The management application will be the heart of the on-line enterprise, because without management the on-line enterprise will be impossible.

### Conclusion

In response to growing global competition and the need to improve productivity, virtually all organizations will use on-line technology to become faster and more effective at what they do. By the year 2001, on-line technology will likely be as pervasive as the telephone is today. In fact, those organizations that fail to become on-line enterprises will not be in business by that time.

One might ask what on-line technology will look like in the year 2001. The most likely answer is that, like the telephone system, it will be an invisible part of our infrastructure. We will rely on the services provided by it, but it will be invisible, tucked away in a closet much as telephone equipment is today. Today the telephone allows us to access a vast and complex network of switches, wires, optical fibers, transoceanic cables, satellites, and microwave links that allow us to communicate instantly with many of the earth's inhabitants. Tomorrow, powerful personal computers capable of voice recognition and multimedia presentation will provide easy, instant access to vast amounts of information and many new services. The California Highway Patrol officer is not likely to know or care about the powerful on-line system that provides information to the computer in his car, but he will appreciate the information that will allow him to do a better job and work in a safer environment.

Therefore it seems likely that by the year 2001 on-line systems will be an invisible but important part of the infrastructure of most organizations. Its main role will be to make possible a higher level of service. Instant access to current information will allow managers to make faster decisions, because they won't need to spend time gathering information. At the same time, improvements in technology will allow transactions to be captured at the source without being processed by a data-entry clerk. For example, it is common today to fill out forms when applying for membership in an organization. These forms are normally entered into a computer by a clerk. By the year 2001, the applicant will enter this information by conversing with his personal computer, which will simultaneously communicate with the organization's on-line system. The results will be lower costs and more accurate data.

Thus, on-line technology will have a profound effect on organizations and on the way they do business into the next century. It will enable organizations to improve the level of service they provide while improving productivity. The services offered will be much more people-oriented as rigid written forms give way to voice input and as multiple inconsistent databases give way to a single consistent view of distributed data.

### References and Notes

1. George Stalk, "Time—The Next Source of Competitive Advantage," *Harvard Business Review*, July-Aug. 1988, pages 41–51.

2. Peter Drucker, *The New Realities*, Harper & Row, 1989.

3. Lynda Applegate, James Cash, and D. Quinn Mills, "Information Technology and Tomorrow's Manager," *Harvard Business Review*, Nov.-Dec. 1988, pages 128–136.

4. Minimizing the effects of cache misses on overall processor performance has been the subject of countless articles over the last two decades. These effects can be illustrated with the following model, which assumes the following:

• a main memory-access time of 100 nsec and one instruction execution per cache cycle

• a 32-bit cache memory data bus width and a microprocessor that can execute up to 3 times faster than the cache access frequency

• instruction and data cache hit ratios of 95% and 97%, respectively

• that every doubling of the number of instructions in the cache halves the cache miss ratio

• that three cycles are required to determine that a cache miss has occurred

• a write-through data cache with a distribution of 20% data writes and 80% data reads.

The execution of 100 instructions requires 100 Instruction Fetches + Instruction Cache Miss cycles + Data Cache Miss cycles. Therefore,

Total cache cycles to execute 100 instructions

$$= 100 + 5(3 + \text{MF/CF}) + 3(3 + 1.2*\text{MF/CF}) \text{ for 32-bit instructions}$$

= 50 + 2.5(3 + MF/CF) + 3(3 + 1.2*MF/CF) for 16-bit instructions

= 40 + 2(3 + MF/CF) + 3(3 + 1.2*MF/CF) for 12-bit instructions,

where MF is the main memory cycle time and CF the cache cycle time.

5. Availability is sometimes defined algebraically as equal to Mean Time Before Failure minus Mean Time To Repair divided by Mean Time Before Failure (A = (MTBF-MTTR)/MTBF).

6. Ball and Bartlett, "OLTP Systems Need to be Fault Tolerant," *Computer Technology Review*, March 1989, page 27.

# *Knowing What Is Known*

# 7

## The Civilizing Currency: Documents and Their Revolutionary Technologies

*Roger E. Levien*

The invention of writing and of a convenient system of records on paper has had a greater influence in uplifting the human race than any other intellectual achievement in the career of man.
*James Breasted*, The Conquest of Civilization

About 5000 years ago, in the rich flood plain between the Tigris and Euphrates rivers, an agricultural tribe, the Sumerians, began inscribing wedge-shaped symbols on clay tablets to record grain inventories and taxes. By taking that apparently simple step—the making of cuneiform marks on clay—the Sumerians set mankind on a profoundly different evolutionary path. For those first inscriptions marked the invention of writing and of the recording of information in documents, and gave mankind its first tools for the accurate cumulation and transmission of human knowledge across space and through time. That capacity to transmit learning from one generation or community to another became the motive force for the spiraling growth of civilization that has brought us from localized agricultural tribes to a global information society in five millennia.

Ever since the Sumerians' time, documents have been the civilizing currency of societies. Their creation and circulation has brought continuity and coherence to societies and cultures, initially by serving commerce and governance and then by disseminating religious ideas and practices. Letters, records, and books have stored and disseminated information; bills, contracts, and constitutions have empowered or constrained action at a distance. Even today, in an age of transient media, documents remain the preferred medium for conveying lasting knowledge and authorizing action in civilized societies.

Documents and civilizations have evolved together, with changes in one setting in motion changes in the other. The concept of the document has changed to include paper and electronic forms, and the technology for creating, reproducing, storing, and distributing documents has advanced from sticks and storage jars to workstations and magnetic disks. The evolution has not stopped; in fact, the pace of change is even faster now. The documents of the future are sure to differ substantially from current ones, with correspondingly great effects on society.

## Some Definitions

### Document

To many people, the term *document* calls forth the image of a formal or legal paper, such as a contract, a will, or a statute, possibly inscribed on vellum and then sealed, rolled, and tied with a silk ribbon. Although Webster's New Collegiate Dictionary admits of that usage, it also provides the following definition: "a material substance (as a coin or stone) having on it a representation of the thoughts of men by means of some conventional mark or symbol." That would clearly admit Sumerian clay tablets, Egyptian papyri, Chinese silk scrolls, and contemporary ink-marked paper into the category of documents; however, it is somewhat less helpful as we contemplate a present and a future in which the "thoughts of men" are increasingly represented on electronic media.

With that in mind, we at the Xerox Corporation have chosen to use a different and broader definition of *document*. We say that a document is "recorded information structured for human comprehension." This definition admits of both paper (substantial) and electronic (insubstantial) documents. We also say that every document has "an author (or authors) who organizes the information in order to inform or influence an audience."

By including the requirement that the information be recorded, we exclude transient conversations, presentations, and performances from the category of documents, even though they are structured for human understanding. By emphasizing "structured for human comprehension," we distinguish between documents and data, since data are structured for convenient processing rather than for immediate understanding.

In addition, data do not (at least, should not) have an author; they arise from phenomena (such as a sale, a birth, a radioactive decay, or the success or failure of some treatment). Nor do data have an audience; rather, there is an analyst who analyzes the data to gain understanding that is often later conveyed in a document.

Every document is recorded on some *medium*. The Sumerians used clay tablets; we use paper. But papyrus, vellum, stone, wood, and cloth have served; film and other plastics are used currently; and various electronic media are gaining widespread application. Although the concept of document is independent of the recording medium, the associated technologies are not, as we shall see below.

Our definition includes some objects that many of us would not be accustomed to consider as documents. For example, a theatrical film or its recording on videotape is a document according to our definition, as is an audio tape recording of a conversation or a performance that in its unrecorded form we would not consider to be a document. It might be convenient to specify the definition of *document* further, so as to eliminate this vast enlargement of the domain. However, as we move on to discuss the future of the document, we shall see that the primarily textual documents that most of us associate with the term will merge and interpenetrate the worlds of audio and visual documents in such a way that any definition that excluded sound and moving images would also exclude many text-rich future documents.

Despite the generality of the term as we have defined it, we shall confine our consideration of documents and their associated technologies to those types that currently and in the past have been principally textual (although these might interpose figures, tables, and graphics). Furthermore, as we move from the past into the present and the future, we shall concentrate on documents used within or as part of the operation of organizations; we will not address *public documents* such as books, magazines, and newspapers. This means that, for the most part, we shall be dealing with *private documents* that are used in business—letters, memoranda, forms, reports, manuals, catalogs, brochures, contracts, bills, and claims—and that, generally speaking, are dealt with in offices. Whereas public documents are available through commercial channels to anyone willing to pay, private documents are normally distributed, at no charge, only to specific people.

Whatever definition one favors, there is a basic truth about documents: they have played a central role in civilized societies as a repository of knowledge, as a conveyor of information and affect, and as an empowerer of action. They have served as the principal information currency of all civilizations.

### Document Processing

No matter the medium in which it is recorded, there are six basic processes that can be carried out on a document.

• The *creation* of the document embodies the acts of authoring and editing the document's content and form, and its initial recording on the document medium.

• The *reproduction* of the document is the act of recording the document's content on multiple instances of the document medium.

• The *transformation* of the document changes its content, its recorded form, or its medium.

• The *display* of the document incorporates both the presentation of the document and the act of reading it.

• The *communication* of the document is the act of transporting it through space.

• The *storage* and the *retrieval* of the document can be thought of as the acts of transporting it through time.

The execution of one or more of these six processes on documents in any medium is what we at Xerox define as *document processing*.

In our view, document processing is a subcategory of information processing, just as data processing and image processing are. It differs from them only in the nature of the information that is being processed—documents versus data or images. And it intersects with them when there are data in the document (as in a spreadsheet) or images in the document (as in a catalog or a brochure). There are also overlaps when a document is represented as an image, or when it can be treated as a sequence of bits (as though it consisted of data) in communication or storage and retrieval. Despite the intersections, many aspects of document processing are unique to this domain and do not reduce to simple applications of either data processing or image processing.

### The Technology of Document Processing

Associated with each document medium is a set of technologies for carrying out the basic document processes. The Sumerians used a wedge-shaped stick to create documents on damp clay tablets; reproduced them in the same way; put them in clay "envelopes" for manual communication; stored them in clay jars, grouped or arranged on shelves to facilitate retrieval; performed any transformations manually; and displayed them simply by making them available for reading (Chiera 1938; Walker 1987; Steinmann 1988). Today, of course, we have a vast array of document-processing technologies for both the paper and the electronic media: pens and pencils, typewriters, printing presses, copiers, mail systems, filing cabinets, word processors, printers, facsimile devices, magnetic and optical disk files, and workstations with associated software.

### Document Processing: The Past

For 4500 years, the metronome of innovation beat slowly. The Sumerian system of cuneiform-inscribed clay evolved toward pigment on papyrus or parchment; but the writing and reading of documents remained the province of the privileged: priests, princes, warriors, and merchants. What advances in technology there were primarily facilitated the processes of document creation and display, reaching their zenith in the techniques that produced the bound and illuminated vellum books of the Middle Ages in Europe. But there was no effective technology for reproducing documents. Consequently, access to books and to the knowledge in them was limited to those with access to the libraries in the monasteries and universities, and then only to those documents that were held locally or within traveling distance. The inherent capacity of documents to cumulate a society's knowledge and circulate it over space and time to other societies was sharply constrained—as, consequently, was the pace of societal development. The constraining factor was the absence of an effective technology for document reproduction.

Gradually, the pace of innovation picked up—first in Asia, where paper and its printing from wood and cast-metal blocks were developed, then in Europe, where wood-block printing evolved as the Dark Ages passed into the Middle Ages.

### The Printing Press

In the middle of the 15th century, a series of small technological advances in metallurgy, metalworking, inks, and paper were brought together by Johannes Gutenberg in the invention of the *movable-type printing press,* the vehicle that enabled the modern era to begin (Usher 1954).

With the arrival of the printing press, the dikes holding back the flow of information broke. The great increase in the circulation of knowledge stimulated the generation of additional knowledge in an explosion that echoes to this day. By democratizing access to recorded information, the printing press set in motion the spread of literacy and education, literature and the arts, science and technology, and commerce and industry that led to the industrial revolution and the creation of democratic governments serving at the will of an informed populace.

Elizabeth Eisenstein describes how print "made the Italian Renaissance a permanent European Renaissance, implemented the Protestant Reformation and reoriented Catholic religious practice, affected the development of modern capitalism, implemented Western European exploration of the globe, changed family life and politics, diffused knowledge as never before, made universal literacy a serious objective, made possible the rise of modern sciences, and otherwise altered social and intellectual life" (Ong 1982).

In a phenomenon that has been repeated with each development of a new document medium, it took another half-century after Gutenberg invented the technology of the printing press before the new expressive opportunities it afforded were recognized and the printed book took its modern form. Gutenberg's goal had been to match the highest art of the contemporary illuminated manuscript. His bible brilliantly accomplished this, but it lacked "page numbers, a consistent grammar and punctuation, a title page and a type face designed for rapid reading." As the printing press spread throughout Europe, other printers innovated. "By 1500, it was possible for the Venetian printer, Aldus Manutius, to combine all the pieces into a work similar to a modern book." (Saffo 1988)

### The Typewriter

It was another 400 years before the next key invention in document-processing technology—the typewriter—appeared. Although a writing machine with raised letters was patented in England in 1714, and over 100 other designs were developed during the next century and a half, the first practical typewriter was not brought to market until 1874. Built by the American inventor Christopher Sholes, it was a response to pressures on information flow arising from the expansion of commerce during the industrial revolution (Monaco 1988).

Although public documents could be created, reproduced, and communicated efficiently owing to the printing press, the growing number of private commercial documents still relied upon a scribal technology. Consequently, the same constraints on the circulation of information that had held back the general society through the Middle Ages were limiting the development of commerce in the beginning of the industrial era.

The typewriter primarily sped up the *creation* of individual documents (rather than their reproduction, as the printing press had done). But the first need of the large modern organization was to create the many different documents required to coordinate its geographically distributed operations. Sholes' "Remington No. 1" and its successors, in concert with the telegraph and then the telephone, met that need quite well.

The typewriter, like the printing press, made possible new forms of documents. The genesis of the memorandum is not as well documented or as celebrated as the genesis of the book; however, it appears almost certain that the typewriter, consorting with bureaucracy, is its parent. It is distinct in its form and style from the traditional external letter, conveying information or orders directly affecting the operation of the organization and often having multiple addresses.

By serving as technology platforms for the bureaucracies that grew up to manage the new organizations and agencies, the typewriter, the telegraph, and the telephone had a profound influence on the development of modern societies in America and Europe. (The situation was somewhat different in Japan and China, where ideographic languages did not lend themselves to convenient mechanical typewriting.)

The typewriter had another profound social effect, as well: the increased demand for workers to be "typewriters" (later "typists") brought women into the office work force and led to the introduction of the female secretary.

### The Plain-Paper Copier

Although the typewriter was a significant step forward in document creation, its proliferation and the growth of organizations built up a latent demand for the reproduction of private documents that neither carbon paper nor coated-paper copiers were able to satisfy. Then, less than 100 years after the typewriter's birth, the third key document-processing invention occurred. In 1938 Chester Carlson, working virtually alone, conceived and demonstrated electrostatic copying, which was brought to practical product form in 1959 as the *plain-paper copier* by the Battelle Memorial Institute and the Haloid Corporation (Golembeski 1989). (The Haloid Corporation later became Xerox.)

It may seem surprising, in retrospect, that more than 20 years passed between the invention of xerography and the introduction of the first commercial plain-paper copier, the Xerox 914. But a number of major companies rejected the invention as unnecessary or uneconomical before Haloid grasped the opportunity. Those companies did not appreciate what becomes clear when we examine the situation from the perspective of document processing: only when the copier became available could the same widespread information circulation be achieved for private documents that the printing press enabled for public documents. In modern organizations, confidential or limited-circulation documents are vastly outnumbered by documents that require distribution in more copies than can be produced by carbon paper, and by documents that undergo secondary and tertiary distribution through the copying of copies. Indeed, the copier did for organizations what the printing press had done for societies: it democratized the access to information. (Until recently there were only 30,000 copiers in the Soviet Union, every one of them under lock and key.)

With the typewriter and the copier, everyone could be a publisher and everyone could be in the audience. The circulation of documents within organizations grew from a stream to a river, conveying the information and the orders needed to coordinate activities in the numerous dispersed units of the modern organization.

### Document Processing in the Paper Domain

The printing press, the typewriter, and the copier comprised the "modern" technology of document processing in organizations through the 1960s. With the file cabinet and the mail system, they provided a complete system for the creation, reproduction, storage, retrieval, and communication of documents recorded on paper. For all intents and purposes, when "document" was said, paper was meant. We shall say, then, that document processing was carried out in the "paper domain," and that for private documents the key technology for creation was the typewriter, that for storage was the file cabinet, that for communication was the mail system, and that for reproduction was the copier.

### Document Processing: The Present

In the 1960s, the widespread commercial use of the digital computer set in motion the next profound series of changes in the technology of document processing.

Early experiments and some practical applications of the large central computer for the storage and searching of computer-based document indexes and for the composition of type demonstrated the potential of that data-processing tool for manipulating text documents as well as numerical data.

### The Word Processor: Inventing the Electronic Document

The real impact of the computer on document processing did not occur until the cost of digital computing had been reduced to the point at which digital document processing could occur on the desk. That point was reached with the invention of the word processor in the 1970s.

IBM first used the term "word processor" with reference to its Magnetic Tape Selectric Typewriter, an electric typewriter that could record a document on magnetic tape and then play it back, pausing for the insertion of things such as individual addresses and salutations. The technology evolved through special-purpose digital devices dedicated entirely to the creation and revision of complex documents to its current form, in which personal computers employing a wide range of word-processing programs achieve the same or even greater capabilities.

The word processor was an important advance in the specific technology of document creation, overcoming many of the annoying inefficiencies of authoring and revising a document on a typewriter. But the truly revolutionary advance brought about by the word processor was the practical and widespread use of a new form of document: the *electronic document.*

### Document Processing in the Electronic Domain

By recording documents on electronic media and displaying them on cathode-ray tubes, the word processor opened a new medium to documents—the *electronic domain*—and broke the inevitable linkage between document and paper. In the electronic domain, documents could be stored on magnetic media and accessed in milliseconds (rather than in file cabinets, where access took minutes), and they could be mailed electronically over voice or data networks, reaching their destinations in seconds instead of days. Various transformations, such as alphabetical sorting or spell-checking, could be accomplished by special-purpose computer programs. Reproduction was easy, demanding no more than transmission of the same electronic message multiple times or simple electronic copying from one storage medium to another.

Yet, despite the convenience of the electronic document for creation, storage, and communication, its human users still demanded paper versions. Paper documents remained easier to retrieve from manual files, were necessary when the recipients were not electronically connected, were essential for authentication by a signature or a seal, and—perhaps most important—were far more portable and more convenient to read.

### The Printer

As a bridge between the electronic and paper versions of a document, word processors had printers. Initially, these were formed-character impact printers, similar to typewriters. Through the use of these printers, electronic documents served as the masters for paper documents, thereby increasing their proliferation. Document processing in organizations now operated in two domains, the paper and the electronic, with the printer serving as a one-way bridge from the electronic to the paper domain. By far the largest amount of information (by some estimates more than 90 percent even now) remained in the paper domain—inaccessible from the electronic domain because the "bridge" carried traffic in only one direction.

### The Facsimile Machine: Opening A New Electronic Domain

In parallel with these developments, another key technology was being introduced into organizations, with the goal of enhancing the communication of paper documents over long distances. The facsimile machine, which originated in 1964 with Xerox's introduction of long-distance xerography, also employed the electronic medium to perform its function, but in a way that was fundamentally different from the way of the word processor. Instead of representing the document in encoded form, with each character having a unique byte code that could be interpreted by the word processor as a letter or a number for sorting or arithmetic calculation, the facsimile machine represented the document as a pattern of black and white spots (or *pixels*—for picture elements) corresponding to the pattern of characters and images on the page. It captured that pattern by passing the document over a light-sensitive image bar and sending the resultant sequence of electrical pulses, after compression, over a telephone line to a receiving printer that marked the same pattern on a sheet of sensitized paper.

### The Electronic Content Domain and the Electronic Image Domain

To distinguish an electronic document produced in encoded form by a word processor from one produced in image form by a facsimile machine, we shall call the former "a document in the *electronic content domain*" and the latter "a document in the *electronic image domain*."

With the facsimile machine, documents that could not be readily entered into the electronic content domain could be effectively and speedily communicated in the electronic image domain. This was of particular importance in Japan, where the ideographic language was not easily adaptable to keyboard entry. The facsimile's use grew much more rapidly in Japan than in the United States because it was the only means available for communicating documents electronically. Recent advances in electronics have made the keyboard a document-creation tool in Japan and other nations with ideographic languages, but the use of facsimile machines continues to grow in those countries as well.

In the electronic image domain, existing paper documents need not be re-keyboarded for entry; however, far more storage, communication, and computation capacity is required for a text document than in the electronic content domain.

In the content domain, each character is typically represented by a byte (eight bits), the average word by six bytes (including spaces and punctuation), and the average typeset page of 500–1000 words by 3000–6000 bytes or 24,000–48,000 bits.

In the image domain, a page is represented by a pattern of pixels whose number depends upon the resolution selected. The practical minimum is about 200 pixels or spots per inch (spi) horizontally and vertically; the current practical maximum is about 600 spi.

At 200 spi, 40,000 pixels are required for each square inch; at 600 spi, 360,000 (9 times as many) are required. A typical 8½ × 11-inch page comprises about 100 square inches of text. Scanned at 200 spi, that would require 4 megapixels; at 600 spi, it would require 36 megapixels.

An intermediate resolution, 300 spi, is in widespread use at present; it would require 9 megapixels. For most current applications, each pixel is represented as one bit; this means that it can be either black or white. When a color or gray-scale image is stored, more bits must be used to encode the various color possibilities or the shades of gray for each pixel, further increasing the number of bits required for each page.

Assuming for the moment that one bit is used to encode each pixel, a typical page in the image domain could require about 200 times as many bits for its representation as one in the content domain. Various compression schemes can be used, however, to remove white space and non-informative matter from a page image, reducing the required number of bits by a factor of 10 or more. Nevertheless, the storage, communication-bandwidth, and processing-speed requirements of document images are at least 20 times those of encoded documents, and the situation worsens when gray-scale or color information is added.

### The Bit-Mapped Printer

Another key technology was introduced at the end of the 1970s when IBM and Xerox offered the first laser xerographic printers. Since then, a succession of dot-matrix, ink-jet, and laser xerographic bit-mapped printers have come onto the scene.

Once again, a seemingly small change in technology opened up a vast new arena of possibilities. These printers could place marks on paper in any arbitrary pattern of black and white dots, at resolutions of up to 400 spots per inch. That made possible the printing of any typeface in a wide range of sizes and orientations, and the printing of arbitrary graphics or images. Until then, computer printers could employ only a finite selection of pre-formed typefaces or a restricted set of geometric shapes.

As was noted above, the function of a printer is to serve as a bridge between the electronic and the paper domain. With bit-mapped printers, the bridge has two spans. The first, which passes from content to image, is the *raster image processor* (RIP, or rasterizer). It transforms the encoded representation of a character in the word processor into the desired pattern of pixel marks, depicting the correct typefont, style, size, and orientation. The second, which passes from the image to the paper, is the *marking engine* (marker), which places the desired pattern of pixel marks on paper.

### The Personal Workstation

The next key technology for document processing was the personal workstation, developed in the early 1970s at the Xerox Palo Alto Research Center. Workstations now range from simple personal computers to high-powered professional models. What these tools have in common is the ability to execute a wide range of application software. With the appropriate program, they become word processors—indeed, as was noted above, they have largely replaced special-purpose word processors. Personal workstations have also led to the creation of software to perform additional document-processing functions.

***Electronic Mail***   Electronic mail has become a widely available function on digital computer networks, enabled both by workstation software and by software on various coordinating network minicomputers or mainframes. The mailed documents, which are in the electronic content domain, can be sent to an individual or to a large distribution list with little difference in effort, further speeding up the circulation of information in organizations. Although the early electronic mail systems were quite limited in regard to the formats of the documents they could transmit, some current systems can distribute documents of very high quality.

***Desktop Publishing***   With the representational power of bit-mapped printers and the capabilities of new software for document composition, it has become possible to compose on a desktop workstation documents employing multiple typefaces, multiple columnar structures, embedded graphics and images, and other characteristics that had hitherto been in the realm of public media or expensive commercial preparation. Whereas word processing gave the individual in an organization a fair degree of expressive power, its range in comparison with what is possible with so-called *desktop publishing systems* is like the difference between a harpsichord and a symphony orchestra. Not surprisingly, placing that much creative power in the hands of untrained users does not automatically lead to its effective use; the need for training and fail-safe mechanisms is great.

***Document Bases***   The recording of electronic-content documents on magnetic media opens the possibility of their compact and convenient storage, the precise and convenient selection of desired documents according to complex criteria, and ready and rapid access to those that have been selected. The increasing capacity of storage media, both magnetic and optical, means that hundreds of thousands of pages of documents can be readily available on the desk. By analogy with databases, we shall refer to these as *document bases.*

The issue with a document base is not the storage of documents but rather their retrieval. There are essentially three ways to characterize a document for retrieval. The first is to describe it by its bibliographic data, as in a library card catalog: author, date, length, where published, and subject matter according to some standard taxonomy. The second is to describe it by the words that appear in its text (ignoring common articles, conjunctions, prepositions, and so on). The third is to describe it by its context, such as its linkage to other documents, its physical location, or some aspect of its creation, receipt, or use (such as the person who provided it, the review that mentioned it, or the letter that contained it). Libraries use bibliographic access, but most offices function according to some combination of a subject taxonomy and a variant of context, usually contained in the mind of a secretary or an assistant and not committed to paper. The most common electronic retrieval aids for document bases combine full-text search for selected words with bibliographic information. Very swift searching of the full texts of even large document bases finds those documents that contain some specified combination of words. Much remains to be done to achieve effective retrieval of documents from increasingly large document bases.

### The Scanner

The latest key technology in document processing—just coming into use—is the general-purpose *scanner,* the function of which is to translate paper documents into electronic pixel images. The scanner is a bridge between the paper domain and the electronic image domain. It operates in the opposite direction from the marking engine, which takes an electronic image and creates it on paper. Scanners are part of every facsimile machine, but they generally have not been available separately. The advent of desktop publishing systems has been a major recent impetus for the more widespread use of scanners, because those devices make it possible to include photographic-quality gray-scale images in the documents produced, creating a demand for the inclusion of images taken from paper.

The combination of the scanner and very-high-capacity optical storage media opens the possibility of storing documents as electronic images rather than as electronic content. Although that precludes access to a document by its content, it offers a means for compact and accessible storage of paper documents. The paper contents of a five-drawer file cabinet can be represented on one optical disk. If each document is given an identifying number, a separate index can be constructed containing bibliographic and context information to make access convenient. For many applications where content retrieval is not important, such document image files are likely to replace large paper files because of the advantages of speedy access by several simultaneous users.

The electronic image domain differs from the electronic content domain in the manner in which the basic document processes are executed. In the image domain, one creates a document through drawing, commonly using a mouse or some other pointing device and drawing software (such as that for creating charts, graphs, freehand illustrations, and engineering drawings). By analogy to "word processors," these programs might be called "pixel processors." Communication occurs via the facsimile device; storage and retrieval employ very-high-capacity optical or magnetic media. Transformations, such as size changes or contrast shifts, are performed by software. The document image requires high-resolution bit-mapped video terminals for display and reading.

### The Recognizer

The scanner, in performing the inverse function of the marking engine, serves as one half of the bridge between the paper and electronic domains. To complete the bridge from paper to electronic documents requires the opposite of the rasterizer, to take the pixel image pattern, identify the character it represents, and transform it into an encoded form in the electronic content domain. The device that does this is called a *recognizer*.

A number of acceptable character recognizers currently on the market recognize the pattern of individual alphabetic characters in a wide range of typefaces with high, although not perfect, accuracy. To do the job in a complete way—providing a full bridge from paper to electronic content—would require that the recognizer also be able to identify font styles and sizes, formats, graphics, and pictorial images (which would not be subject to further recognition efforts) and to recreate an encoded representation of the document (with embedded pictorial images) that could be used by another printer to produce an exact replica of it. Such *document recognizers* do not exist at this moment.

### The Three Domains

At the present stage in the development of the document and of document-processing technology, private documents can exist in three domains: paper, electronic image, and electronic content. Each of the six basic processes can be executed in each domain, and technology exists that—with some limitations—allows bridges between the domains to be built. This state is represented in figure 1. Yet the completeness and symmetry of the picture is somewhat misleading. Most information in most organizations is still in the paper domain. The electronic content domain is widely used for document creation and electronic mail, and its use for storage and retrieval is increasing. But technology is enabling the electronic image domain to grow beyond the facsimile's use for communication. The role of this

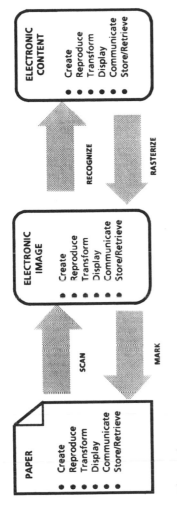

**Figure 1**
The three domains of document processing and the bridges between them.

domain for storage and retrieval of documents can be expected to expand significantly, especially as a replacement for large paper or microfilm files.

The further evolution of document-processing technology will be driven in large measure by electronic documents, which are opening new possibilities for use. The 1970s and the 1980s saw the birth of the electronic document, but we have just started to realize and exploit its potential. The paper document remains the preeminent repository of knowledge and the preferred medium for authorizing action in all civilized societies. However, looking ahead to the last decade of the 20th century and the early decades of the 21st, we can see a different picture forming. In the coming decades, electronic-document technology and the electronic document promise to become preeminent and, in doing so, to change profoundly the societies they serve.

## Document Processing: The Future

During the next two decades the metronome of change in document processing technology will tick ever faster, driven by the continuing revolution in digital computing and communication technologies and by steady advances in the technologies of paper handling, marking, and finishing.

### Information Technology and Architecture

The microelectronics revolution will progress unabated, driving microprocessor speeds into the hundreds of millions of operations per second, workstation random-access memory capacities into the hundreds of megabytes, and local mass storage capacities into the tens of gigabytes. At the same time, digital communication capacities will expand to a hundred megabits per second locally and tens of megabits per second over longer distances. High-resolution flat panel displays will become common, and high-quality graphical user interfaces will be ubiquitous.

Printers will be capable of very-high-resolution (600 spi and over) printing in black and white and gray-scale and color printing at lower resolutions. They will be available in a wide range of sizes and speeds, from less than one page per minute through hundreds of pages per minute. Sorting and finishing (including insertion and binding) will be commonplace, even for small printers.

As a result of these component changes, the basic architecture of digital information systems will move away from the centralized model in which a mainframe or a minicomputer is the controlling element in a hierarchically structured system. Instead, the basic architecture of digital information systems will be *workstation–network–server*; with powerful workstations linked to one another and to shared resources (servers) by high-capacity networks, which will be tied to other networks both collegially and hierarchically. Most document processing will be performed on such architectures, in conjunction with other forms of information processing.

### Features of Document Processing
As a result of these technological and architectural developments, three fundamental features—transparency, decentralization, and modularity—will characterize future document-processing systems.

*Transparency* Documents will move transparently between domains. Paper documents will be readily transformed into electronic documents, and vice versa. That means that documents will be transformed into whatever domain is most convenient or appropriate for the purpose at hand. For instance, many of the existing stocks of paper documents will be moved into the electronic domain for convenient communication and storage.

*Decentralization* Many of the document processes now carried out centrally or departmentally will be available at the desktop or the deskside of the individual user. This distribution of function will be driven by reductions in the cost of basic document-processing technologies, and by the ready availability of high-capacity communications lines. Central facilities will retain their advantage for jobs requiring highly specialized capabilities or skills, or for production in large numbers or very high quality; but many of their previous tasks will occur much closer to the end user.

*Modularity* Basic functional modules will be used separately, linked by the network, or combined locally in various ways to achieve a range of single-function or multi-function devices.

Scanner and printer modules will be combined to perform all the functions now carried out by scanners, printers, copiers, and facsimile devices.

These fundamental features will combine with the directions of information technology and architecture to create three central document-processing technologies for the not-too-distant future: *docugraphic systems, document servers,* and *documentors.* In these systems, paper and electronic documents will coexist. The electronic document will be preferred for creation, transformation, storage and retrieval, and communication, but in the end it will serve as an "electronic master" for the printing of a paper document, which recipients will continue to prefer for convenient reading.

### Docugraphic Systems
A significant result of these technological and architectural changes is that the functions of copying, printing, scanning, and long-distance copying (facsimile transmission), which have heretofore been associated with separate devices and somewhat different technologies, will be carried out by various combinations of modules linked by networks. In other words, they will become functions on or of network systems.

Let us call a network that contains and links scanners and printers a *docugraphic system.* In such a system, the scanning and printing functions will be carried out directly by the corresponding devices. Copying, however, will be accomplished by a scanner followed by a printer in close proximity. Facsimile transmission will employ the same combination, but with the scanner and the printer located at a distance from each other. In addition, local files will be associated with each scanner or printer to hold work in progress and to buffer multiple requests for service.

The copier—which has been a separate box, distinct from the remainder of the information system of the firm—will thus become an intimate part of it. Indeed, it may disappear as a separate product, becoming simply a function carried out by a scanner-printer system.

The digital printers in these docugraphic systems will deal as readily with color and gray-scale images as they do today with black-and-white text. Many printers will be able to handle documents with intermixed black-and-white and color pages almost as efficiently as they handle black-and-white documents. The marking technology that will make this possible will probably involve the direct laying down of ink or toner on paper, rather than the indirect transfer process that underlies today's xerography and ionography.

A wide range of printer capabilities will be connected to the network. Speed distinctions will still govern the location of printers. Very fast printers will produce over 200 prints per minute in black and white on both sides of the page, insert separating or tab pages and covers, and bind the document. Because of their expense, they will be located in central printing facilities in a corporation or a service bureau. Printers producing from 30 to 100 prints per minute are likely to be shared by groups or departments able to support their high capacity. Slower printers, producing from 1 to 30 prints per minute, will be used by individuals or small work groups.

Users will be able to select the printer whose location, speed, quality, finishing features, and availability best match their needs. They will also be able automatically to distribute documents directly to the printers closest to their intended readers.

Similar ranges of quality, features, and locations will characterize the scanners connected to the network. Many of them will be located at workstations, so that users will be able to enter paper documents to be copied via the scanners at their desks and to have them printed at one or more printers that they might designate. For example, if you need thirty copies you might send the document to the local printer down the hall (copying). If you need single copies, you might send it to ten other printers at different locations around the organization (facsimile). Someone with a very large volume of paper documents to scan will take them to a centrally located high-capacity scanner, from which they will be distributed to printers or files on the network.

The high volume of scanned information moving around the network will require very broad transmission bandwidths, as well as very fast processing and very high storage capacity. Scanned images, even after compression, will impose heavy requirements on the corporate information infrastructure.

### Document Servers
The existence of large numbers of documents on the network will bring into existence facilities to carry out specialized func-

tions that could not be so well performed at individual workstations. A wide variety of *document servers* will be available to users at workstations or to other servers on the network. Among those most likely to be widely used are *storage/retrieval servers, communication servers, transformation servers, recognition servers,* and *document managers.*

*Storage/retrieval servers* will manage large document bases, maintaining them on very-high-capacity files (tens or hundreds of gigabytes) associated with very-high-speed files containing the indexes through which they will be accessed. The indexes will allow retrieval to be based on one or more of the document's contents (as represented by combinations of words appearing in it), by its context (as represented by its history or its associations with other documents, events, places, or people), or by its bibliographic description (as given by its author(s), source, date, length, and subject terms). Large capacity and access via content or description will come first; more sophisticated techniques of accessing a document via its associational context will develop over time.

*Communication servers* will enable documents entered via a workstation or a scanner on a local network to be sent to workstations or printers on other networks anywhere around the world (where government policies and communications availability permit), wholly transparently. They will generalize and merge the capabilities of today's electronic mail and facsimile transmission. Documents of any size and any textual or image complexity will be communicable—including compound documents that incorporate audio or video. A key function of the server will be the maintenance of directories (and directories of directories) that will enable us to reach a person by a simple address, without having to specify the precise path.

*Transformation servers* will provide the ability to move seamlessly among documents in a variety of formats. (The use of different hardware and software systems will continue.) The communication server will most likely call upon the transformation server to make any necessary format conversions before forwarding a received message to the recipient workstation or printer. Format conversions may develop over the years into linguistic conversions; acceptable translations between natural languages might become feasible. Yet another form of transformation is the translation from text to speech. Speech synthesizers already exist and are widely used in commercial systems to respond, for example, to telephone queries. This technology will surely advance in its mimicry of human voice quality, and will simultaneously be reduced in price to the point that there will be no difficulty in having a textual document "read" to you by your workstation if you prefer rather than having it presented visually.

*Recognition servers* are likely to be closely associated with scanners, just as raster image processors are associated with printers at present, although (like RIPs) the recognition servers will also be locatable in workstations as software or special-purpose hardware. The function of the recognition server will be to take a raster image of a page and translate it into an encoded representation capable of driving a printer or a display to re-create an exact image of the document. Such recognizers will operate with very high accuracy and speed, making truly transparent movement from the paper domain to the electronic content domain practicable. Thus, mail or old documents or clippings or journal articles will be capturable for filing and content access, for efficient communication, or for ready incorporation into other documents. A related kind of recognition server will translate the audio patterns of recorded speech (comparable to the visual patterns of imaged documents) into recognized words, storable in the same text encodings as visual documents.

*Document managers* will be yet another kind of server function. These will keep track of the location and flow of documents within a network of document users. For example, in an environment in which documents are processed in order to complete a transaction, such as an insurance claim or the auditing of an income-tax return, the document manager will handle the sequencing and recording of the work flow as the appropriate documents move through the necessary steps. Alternately, where complex documents are being created, as in the publishing of a technical manual or a catalog, the document manager will store the components of the document, keep track of various versions and drafts, and monitor the document's progress against a specified schedule. Yet another role for the document manager will be to manage the distribution of documents to appropriate recipients. By storing an interest profile for each "user" and comparing documents entering a document file or passing through a mail system with that profile, the manager can select, categorize, and even prioritize documents for the user (Malone et al. 1987). The software resident in such a document manager need not be rooted in the local hardware; in the future, a copy of the software might be given a description of a specific immediate information need and then "mailed" to a number of document bases located at various places on the network, where it would be "received" and enabled to perform a local document search according to the specified requirements, mailing back the retrieved documents meeting the specified need.

### Documentors

The third major document-processing technology of the future will be new forms of document-creation tools, varying widely in size, capability, and user interface, which we shall call *documentors*. These are really workstations with software and other features designed for document creation. For the next decade or so, they will continue to resemble contemporary workstations and personal computers; but as they get lighter and as their displays improve in quality, the result—by the turn of the century—will be a tool we at Xerox call *Dynapaper*. When it becomes available, it will pose, for the first time in 500 years, a challenge to paper as the preferred medium for document display and reading.

***Personal Dynapaper***   Paper remains in use in the electronic age because its advantages in readability, portability, cost, and familiarity outweigh the dynamism, flexibility, multimodality, and interactivity of the electronic medium. However, as electronic technology advances, its capabilities will come to match those of paper. In the early 1970s, Alan Kay, then at Xerox Palo Alto Research Center, conceived of Dynabook, a book-size computer that would have many of the capabilities of today's laptop computers (Kay and Goldberg 1977). Although its imminent existence will certainly challenge many of the uses of paper, technology is pointing in a direction that offers an even greater challenge. By the end of the century, we shall see an electronic device about the size of a thin stack of 8½ × 11-inch paper— *Dynapaper*. Almost all of its surface will be a flat panel display with a pressure- or light-sensitive transparent covering. A rigid bar along one edge will contain very-high-capacity memory chips, a very-high-speed processor, and a battery. The display will be paper-white and will have sufficient resolution to display high-quality text with multiple fonts and gray-scale images, as well as color. The memory will be able to contain at least 100 such pages. Documents will be entered into the Dynapaper via a connection to an electronic network or file and then carried anywhere that paper could go—on an airplane, to a comfortable chair, to bed, or to a conference room. When the pressure- or light-sensitive surface of the Dynapaper is written upon with a stylus, programs in the embedded processor will recognize the hand-printed characters and immediately display them in a formal typeface. Thus, Dynapaper will function both as a notepad and as a workstation.

***Group Dynapaper***   Another form of documentor that will evolve in the future will serve people who are working together. It will be a wall-mounted Dynapaper-like device, the size of a whiteboard, linked to personal Dynapaper tablets in front of each place at a conference table. Access to the main screen will be gained, under the leader's control, by any one or more of the participants through their tablets. The group's work will be cumulated on the screen and distributed to each participant, who will be able to take it back to his own office for further work or to send it on the network for additional contributions.

These two extremes—personal Dynapaper and group Dynapaper—both represent the consequences of trends in which the processor and memory components of the workstation will be shrinking to near invisibility, leaving the display and the input device as the largest and most visible components. As displays become thinner and as other means develop to complement the keyboard and the mouse, workstations will take a variety of forms determined primarily by their display and entry needs. Thus, we can expect to see documentors embedded in desks, briefcases, and conference tables.

### The Future of the Document

This series of technological innovations, and the rapid pace of evolution in the underlying digital and electronic worlds, will lead to the creation of entirely new forms of documents, capable of substantial interaction with their environments and their audiences. Electronic documents will become a new medium, opening possibilities for content and structure that have been impossible in the paper medium and demanding new technologies—such as Dynapaper—for their creation, display, and publication.

***Compound Documents***    An electronic document can incorporate not only text and graphical images, but also audio or moving video. At present, for instance, it is possible to use a telephone system to send a voice message that will be held in an electronic mailbox until it is retrieved by its intended recipient. Such messages are *electronic voice documents,* and the systems are called *voice mail systems.* It is also possible to embed a voice message within an electronic text document, by recording it digitally and including it in the encoded document. This capability is called *voice annotation,* because a clear use for it is to enable the recipient of a document to insert comments and reactions as annotations and return the document to its sender, who can then listen to the annotations. What is currently true of audio or voice should in the not-so-distant future also be true of video; that is, it should be possible to send recorded video messages (*electronic video documents*) over electronic networks (*video mail systems*). And, by the same technology, it should also be possible to annotate documents with video images. A document that incorporates not only text and graphics but also still images, audio, and video is called a *compound document.* These currently exist in their full form only in experimental systems. However, they can be expected to form the core of the document of the future.

***Active Documents***    Another way in which the electronic document can extend its capabilities beyond the paper document is through computational activity. Unlike a paper document, which is inherently passive, an electronic document can be given the capacity to respond actively to information from the external world, becoming what we shall call an *active document.* A spreadsheet, for instance, has formulas defining some of its cells in terms of others. If the entry in one cell changes, then many of the others change as a consequence. This concept of one part of a document being dependent (in a defined way) upon changes in other parts can be made quite general, leading to active documents other than spreadsheets. For example, a bill or contract form might include a number of variable entries, some of which would be provided from the outside (such as the name of the customer, the specifics of the order, or the general terms of the contract) and the remainder of which could be automatically calculated or filled in on the basis of pre-specified routines. This capability can also be extended to sets of documents, which can be interlinked so that entries or changes in one will generate appropriate changes or entries in the others.

***Expert Documents***    When the rules used to fill in the dependent entries are based neither on strict arithmetic nor on clear procedures but rather on the "expertise" of a specialist, the active document embodies "artificial intelligence" and might be called an *expert document.*

***Interactive Documents***    Another way in which an electronic document differs from a paper document is through the possibilty of interactivity between the document and its reader. An *interactive document* is one whose content or sequence of presentation can be altered through direct intervention by the reader at the time of reading. For example, if an active document is being read on the display of a personal workstation, the reader may enter some information that will affect the next information presented. Imagine a survey questionnaire, for instance, that is presented by means of a workstation display. As it is answered by the respondent, the nature and the sequence of subsequent questions can be changed to reflect what has been learned. Irrelevant questions can be skipped and others can be personalized. Although many electronic documents intended for "on-line" usage incorporate some degree of interactivity, the potential for such responsiveness is only beginning to be exploited.

The anticipation of these complex documents is not, however, a recent development. In a famous article entitled "As We May Think," published in the *Atlantic Monthly* in July 1945, Vannevar Bush described a device, called the Memex, that would open a world of logically interconnected documents to browsing, associative linking, and interactive retrieval by its users. His ideas were taken up and explored by Douglas Engelbart at the Stanford Research Institute during the early 1960s. One result of Engelbart's work was the digital mouse now widely used for computer interaction (Engelbart and English 1968).

**Hypertext**    In the late 1960s, Theodor Nelson took Bush's concept further, calling attention to the potential offered by the computer and the video display to create a new medium, which he called *hypertext,* that would not follow the linear presentational progression imposed by printed and bound documents (Nelson 1967). Rather, the user would be able to steer his computer display electronically along a course, which he selected to meet his requirements, through a multi-dimensional document space. At each step, choices would be available to dig deeper, or see an example, or examine the evidence, or work a problem, or view a simulation. The repertoire of such choices would depend upon the author's creativity and the availability of software to implement specific options. In the years since Nelson's farsighted anticipation there have been numerous efforts to implement his vision, but only now is the technology becoming widely enough available to begin to bring it into common use (Conklin 1987).

Nelson, who comes from a family active in the worlds of theatre and film, saw that hypertext was not simply a variant of existing documents but an entirely new medium, just as film differs from theatre and as the printed book differs from the vellum manuscript. He pointed out that until D. W. Griffith developed the "grammar" of the film in the mid-teens of this century (varying close-up, medium, and long shots; using flashbacks and jumps in time), motion pictures had been simply filmed stage plays; after that, they developed as a separate medium capable of telling stories or conveying information by means unavailable on the stage. In the same way, the change from linear to multi-dimensional documents, together with activeness, interactivity, and incorporated audio and video, opens an entirely new creative arena that demands its own D. W. Griffith to show the way to exploit its expanded capacity to convey information and emotion.

Insights into what may become possible are provided by some of the initial implementations of hypertext-like capabilities. One variant employs the *notecard metaphor*: the elemental documents are the electronic equivalents of 3 × 5 cards, each of which can be electronically linked to any of the others in expression of certain user-defined relationships. For example, if the cards contain bibliographic information for an article, the linkages may be based on subject, author, date, relevance to specific sections of the article, or location. Then, the author could begin with one card, look at the others on the same subject, examine what else an interesting author had written, go from that to the items relevant to the introductory section, and go back to one of the other items again—all without leaving his screen or doing anything more than selecting a link to follow from those displayed (Halasz 1988; Goodman 1987).

Another variant emphasizes the same multi-dimensional linking property, but takes as the linkable element an audio-visual segment. One dramatic example of this direction of development was a videodisk tour of Aspen, Colorado (prepared by the Architecture Machine Group at MIT), in which the viewer can "steer" a car through the streets of Aspen, and the videodisk will display the view ahead appropriate to the current location, no matter how it was reached. But clearly, the sequence of audio-visual segments can be controlled in any number of different ways, not only via the path of an automobile. For example, in an instructional audio-visual document, the sequence can be selected according to the student's responses; in a dramatic audio-visual document, the sequence can be determined by the viewer's interest; in a reference audio-visual document, the sequence can be selected directly by the user.

The arts of writing and reading traditional linearly organized documents have developed gradually over the centuries, albeit at greater speed in the 500 years since the printing press came into wide use. Hypertext, with its much wider range of choices both for each informational unit (text, graphics, audio, video) and for the range of connections among them, poses a much greater challenge to the author and to the reader. In some cases, the subject matter might lend itself more naturally to nonlinear organization and might suggest readily what the links might be. For example, an encyclopedia comprises distinct segments of information with many linkages among them that are partially specified in each text element and partially captured in a separate index; hypertext implementations will allow each link to be shown and to be followed instantaneously under the reader's control. Textbooks and technical manuals also fall naturally into the hypertext format.

But hypertext has yet to find its Aldus Manutius or its D. W. Griffith. If Nelson's vision is achieved, we can expect to see entirely new kinds of documents, in which authors exploit the

multi-dimensionality and the multi-media characteristics of hypertext to provide new informational and emotional experiences for readers. And we can also expect to see readers
developing new reading skills that will permit them to exploit
the full power of active engagement with the material—and the
shaping of its presentation to their own needs and capacities—
that hypertext offers.

**The Demise of the Paper Document**   None of these new forms
of documents can be adequately represented on paper, which
cannot display audio or video segments, change contents actively or interactively, or readily enable multiple alternative
paths through its contents. Yet constraining their display to
desk-bound workstations is bound to limit their impact. They
demand a device like Dynapaper, and they will flourish only
when such a device becomes available. Reciprocally, when
Dynapaper becomes available, the market for the electronic
document with no paper version is likely to expand rapidly.
Only then would a "paperless" world become feasible and
desirable.

But the organizational and physical infrastructure of public
and private document distribution and acquisition has been
built upon the paper document. For the electronic document
fully to replace the paper document, an infrastructure satisfying the same organizational and physical requirements would
have to be established for electronic documents. How, for example, would the publishing of electronic documents occur?
How would authors be compensated? How would private documents be distributed? Here again, Ted Nelson has been in the
conceptual avant-garde, with a concept called Xanadu.

We may also ask how the individual will build a library of
electronic documents comparable to the library of paper documents one now builds up over a lifetime. Here we may return
to Vannevar Bush's 1945 paper for inspiration: technology
will make possible the creation of a Personal Memex. Each of
these future document technologies may have profound social
consequences.

**Xanadu**   During the 1970s Theodor Nelson and a few colleagues extended their inventive vision from the individual hypertext document to a hypertext literature available through a

"universal electronic publishing system and archive," which they called Xanadu after the idyllic "stately pleasure dome" of Coleridge's poem "Kubla Khan" (Nelson 1987).

To Nelson, a literature is "an ongoing system of interconnecting documents." Each of the documents in a literature may comprise both original content and quotations from or references to short or long segments of the content of other documents. Moreover, each document may exist in a number of dated versions; these must all be retained in the literature, both as a historical record of the document's evolution and because the variant content of any revision may be referenced by other documents. Nevertheless, each document version need exist only once in the literature, because all references to it or its segments in other documents would be implemented as pointers to the original, no matter where it might be located.

In Xanadu, the literature will be stored on distributed document servers linked by networks. Each document will be on only one server, and will be represented by the combination of its unique content and its quotation pointers to other documents on the same or other servers. A book review, for example, would comprise the reviewer's text plus pointers to the text segments in the reviewed document that are included in the review. Since each document may both point to other documents and be pointed to by others, the literature will comprise a complex network of interconnecting documents. Documents on the system will be public, private, or semi-private.

The act of making a document public is *publishing*. It means that the document may be accessed by anyone, and it entails contractual obligations, protections, and financial compensations similar to those in commercial book publishing. The author or owner of a public document will pay for its storage and will receive royalties that are automatically computed every time the document or a segment of it is referred to by a user, whether the reference is direct or through another document. The document, once published, can be neither changed (except by publishing a new version) nor removed (except to lower-cost, slower storage, or by a lengthy process). References to and unlimited quotations from a public document will be encouraged, because the system will ensure that each use of such a reference or quotation generates a royalty payment.

In contrast, a *private* document will be accessible only to the author and to a designated list of associates. Only they will be able to read it or link to it. Should an author wish to make a document available to all while not incurring the obligation to leave it available permanently, he will be able simply to designate it a private document with unrestricted distribution. Nelson calls this *privashing*. No royalty will be received for the use of these *semi-private* documents.

The user will connect to the system from his workstation, paying a fee for its use that will include both a charge for the technical facilities and service and a royalty charge that will be paid to the owners of the public documents accessed. Although the documents the user will access may be physically stored in many different servers on the network, he will be completely unaware of their different locations. High-capacity communication networks and high-speed storage and retrieval of documents and their links will provide transparent and speedy access to documents anywhere on the net.

Nelson and his colleagues have been working on the Xanadu project for many years, achieving implementation of some portions of the system in prototype form. In 1988 a software publisher, Autodesk, acquired a majority interest in Xanadu with the intent of developing the hypermedia document servers that are the key components of what Nelson envisages as the "principal publishing utility of the future." If it fulfills his vision, Xanadu will "provide for the deposit, delivery, and continual revision of linked electronic documents, servicing hundreds of millions of simultaneous users with hypertext, graphics, video, movies and hypermedia" (Nelson 1988).

Although the eventual embodiment of an electronic document publishing utility may not incorporate all the ideas reflected in Xanadu, it will have to address many of the same issues of document linkage, distribution, publication, royalties, and fees. And, most profoundly, the creation of such a utility will open the possibility of publication to anyone who can access the network and pay for storage in one of its document bases, without requiring the approval of a commercial publisher. Although in all free countries there is the irrevocable *right* to publish, this would be a revolutionary democratization of the *capacity* for publication. It would mean that any person could create a document and make it available to anyone on the utility (eventually, virtually everyone, as on the telephone system) at minimal cost and effort. In such a situation, of course, the demand will quickly arise for services that—for a fee—identify documents of quality and value that have been published or "privashed" on the utility. Such services may be the form that today's publishers will take in the future.

***Personal Memex***   The empowerment of the reader to engage actively with documents—facilitated by hypertext, Dynapaper, and vast electronic publishing utilities—should finally bring to pass the grand vision conceived by Vannevar Bush in 1945, but in an even more visionary form. In addition to public Memexes, like Xanadu, accessible via Dynapaper or other workstations from the network, the development of very-high-capacity personal storage media (holding hundreds of gigabytes of documents in the same spatial volume as a single book) should make possible the *Personal Memex*.

The Personal Memex will be a storage accessory, portable and connectable to Dynapaper or other individual workstations or to a network, for the storage or retrieval of documents from electronic publishing systems. The material would be stored in a hypertext format, with some links provided automatically and some created by the reader. Individuals would fill Personal Memexes by adding whatever documents they found interesting or important, defining links to already-stored documents, and adding new linkages as new relationships became evident or relevant. These Personal Memexes would be large enough to contain the documents selected over a number of years, and so would become a repository for the knowledge deemed important by an individual over his educational or professional career, organized in his own way and restructured as his own perceptions and understandings evolve.

Each individual would have the capacity to build his own personal library, a "life file" of all the books, articles, letters, memoranda, songs, pictures, and video segments that had significance for his career or personal life. Any item in that file would be accessible at any time from a personal workstation or Dynapaper.

Thus, in a sense, the last step in the democratization of access to information will have been brought about. As the printing press made it possible for everyone to have his own books, the Personal Memex will make it possible for everyone to build his own vast library and to access each element with the facility of a professional library staff.

### The Impacts on Individuals, Organizations, and Society

As was noted above, there is always a time lag between the introduction of a significant new technology and its integration into human thought and practice.

Although the concepts of written language and the printed book have been thoroughly internalized in almost every culture, the former has been around for more than 5000 years and the latter for 500. Current innovations in document processing may well have an equally profound effect on the evolution of society and human thought, but the new approaches to interacting with information that these technologies will engender are as yet unknown.

Avant-garde works of art violate implicitly held conventions and disrupt standard interpretation, making such works difficult to understand until the broken conventions themselves become stabilized as new conventions. We can consider ourselves today to be in an era of avant-garde technology. Mastery of the new media will be the result of a social process in which both designers and users will contribute to the construction of a new understanding (Brown and Duguid 1989).

The "alphabet" of electronic document technology will one day be as thoroughly internalized as the conventional alphabet—and it will achieve the same sort of transparency that allows people to see through the marks on a page to the world to which the marks refer. At this point, however, predictions of the impact today's document-processing technology will have on future generations can only be of a general nature, based on our observations of the past.

By looking back over the five millennia since the Sumerian invention of writing, we have been able to see the vast civilizing influence that documents and their evolving technology have exercised on human society. At each step along the way, technology has led to a wider and more effective circulation of information, setting in motion profound changes in social organization and accomplishment.

Looking to the future, we can foresee the continued acceleration of society's progress with each increase in the rate of circulation of that civilizing currency, the document.

**Bibliography**

Breasted, James H. 1926. *The Conquest of Civilization.* New York.

Brown, J. S., and Duguid, Paul. 1989. Innovation in the Workplace. Institute for Research on Learning, Palo Alto, California.

Bush, Vannevar. 1945. "As We May Think." *Atlantic Monthly,* July, pp. 101–108.

Chiera, Edward. 1938. *They Wrote on Clay.* University of Chicago Press.

Conklin, Jeff. 1987. "Hypertext: An Introduction and Survey." *IEEE Computer,* September, pp. 17–41.

Eisenstein, Elizabeth L. 1983. *The Printing Revolution in Early Modern Europe.* Cambridge University Press.

Engelbart, Douglas C., and English, William K. 1968. "The Augmented Knowledge Workshop." *AFIPS Conference Proceedings* 33, pp. 395–410.

Golembeski, Dean J. 1989. "Struggling to Become An Inventor." *American Heritage of Invention and Technology,* winter, pp. 9–15.

Goodman, D. 1987. *The Complete HyperCard Handbook.* New York: Bantam.

Halasz, Frank G. 1988. "Reflections on Notecards: Seven Issues for the Next Generation of Hypermedia Systems." *Communications of the ACM* 31, no. 7, pp. 836–852.

Kay, Alan, and Goldberg, Adele. 1977. "Personal Dynamic Media." *Computer,* March, pp. 31–41.

Malone, T. W., et al. 1987. "Intelligent Information-Sharing System." *Communications of the ACM,* May, pp. 390–402.

Monaco, Cynthia. 1988. "The Difficult Birth of the Typewriter." *American Heritage of Invention and Technology* 4, no. 1, pp. 10–21.

Nelson, Theodor H. 1967. "Getting It Out of Our System." In *Information Retrieval: A Critical Review,* G. Schechter, ed. (Washington, D.C.: Thompson).

Nelson, Theodor H. 1987. *Literary Machines,* Version 87.1. San Antonio, Texas: Project Xanadu.

Nelson, Theodor H. 1988. "Managing Immense Storage." *Byte,* January, pp. 225–238.

Ong, Walter J. 1982. *Orality and Literacy: The Technologizing of The Word.* London and New York: Methuen.

Saffo, Paul. 1988. "What's Beyond Paper?" *Personal Computing,* December, pp. 69–74.

Steinmann, Marion. 1988. "'Chicken Scratches' Written in Clay Yield Their Secrets." *Smithsonian* 19, no. 9, pp. 130–141.

Usher, Abbott P. 1954. "The Invention of Printing." In *A History of Mechanical Inventions.* Cambridge, Mass.: Harvard University Press.

Walker, C. B. F. 1987. *Reading the Past: Cuneiform.* London: British Museum Publications.

# 8

## The Multiplier: Future Productivity in the Light of Optical Storage
*Olaf Olafsson*

The information age has introduced an environment in which microelectronic devices can conduct a dialogue, send pictures through the galaxy, instantly perform calculations that used to take weeks, capture three-dimensional images of the insides of our bodies, and play music with laser light. A crucial step in enhancing these achievements is the development of mass storage.

*Mass storage* usually refers to external disk drives, tape systems, and other devices and media that can statically hold more information than fits inside a computer's microchip memory, which is also known as random-access memory (RAM). Spreadsheets, multiple-page documents, engineering drawings, catalogs of parts, and databases of customer records are too big for RAM. Until recently, in fact, bulky records were too big for computer storage at all, and were kept on paper or microfilm.

Information can be retrieved from RAM nearly instantaneously, but RAM is expensive. Mass storage is much less expensive than RAM, but it is slower. Today's mass storage media include, in addition to magnetic media such as the floppy disk, tape, and the hard disk, the *optical disk*. The optical disk has more capacity, and is more versatile and less expensive, than any previous storage device.

Mass storage has always been one of the few fundamental elements in any computer system, along with processors, RAM, and input/output devices such as keyboards, printers, and displays.

Because mass storage is so essential to any computer system, an increase in its capabilities broadens the usefulness of computers and the productivity of their users. What if a disk could hold as much information as a shelf of reference books? What if a single disk could store the voice of an expert mechanic, a set of engineering drawings, and video footage of the component being repaired? What if a person's medical records could be carried on a card or on a disk that would fit easily in a short pocket? What if it were possible to see realistic synthesized pictures of what a newly planted garden would look like in five years, or how a new product would evolve under varying market conditions? What if were possible to access any piece of information in the world in a few minutes?

Computers, and electronic devices of all kinds, process, transmit, capture, recombine, synthesize, display, and sort data. When data are processed intelligently, the product is information, which enhances analysis, planning, and decision making.

Storage capacity eventually places a limit on the amount of information available. Increasing the capacity of mass storage devices makes it possible not only to do bigger things but also to do new and different things. Optical storage can multiply the capabilities of the computer.

There will be a profound effect on the way institutions acquire, process, and disseminate information. Schools use information to stimulate the intellect. Governments use information to stimulate the economy, maintain order, and administer justice. Business uses information to design products and to serve markets. Institutions strive to process and disseminate information at the highest possible speeds. But the amount of information available can become so large that organizing and processing it is an increasingly enormous challenge.

James Burke, who examines the role of science in human affairs, emphasizes that throughout history, whenever new information technologies have radically altered the amount of information available, existing institutions have been reorganized into more complex forms. Indeed, optical storage is already beginning to change our lives and our institutions. In the near future, businesses, schools, governments, and homes may become unrecognizable in the ways they communicate (both internally and with one another), in the complexity of their structure, and in the scope of their responsibilities.

### How Optical Storage Works

Optical storage uses a laser beam focused to a tiny spot to write data on a disk, a tape, or a plastic card. On a rewritable optical disk, data are written magnetically. In other forms of optical storage, actual physical marks are made. In both cases, because the laser spot on the disk is so small—a micron or less in diameter—data can be packed far more densely than on a magnetic disk or tape. A micron is a millionth of a meter. Approximately 100 million of the marks made on an optical disk would fit on a fingernail—give or take a few million. This dense packing of marks or magnetic domains, which we refer to as *high data density,* makes it possible to achieve large storage capacity.

But the small marks alone are not sufficient, as researchers discovered when they began developing optical storage in the 1960s. The smaller the marks, the harder it is to keep a drive head positioned directly over them, especially on a spinning disk. The head is the interface between the physical marks on a disk and the electronics that convert those marks into a signal that carries digital data. The head reflects a laser beam off the disk in order to read the marks on it. If the disk is writable, the head shoots a laser beam at the disk in order to make the marks in the first place.

An optical head consists of a tiny diode laser; a lens; mirrors and other optical components; and *actuators,* which move the head in three directions. Tracks on the disk are spaced 1.6 microns apart, and the head must be moved back and forth laterally to keep the head the proper distance above the disk, so that the laser beam stays in focus on the disk surface. Still other actuators constantly move the head slightly forward and backward, linearly along the track, so that the drive stays synchronized with the marks moving past the head as the disk spins at speeds ranging from about 200 rpm in the case of the inner tracks of a CD-ROM to about 3000 rpm for the fastest commercial optical drives.

The dimensions are so small and the movements so fast that they are unimaginable in terms of direct experience. If the whole apparatus were scaled up to more familiar dimensions, an average optical drive reading a disk would be the equivalent of a Boeing 747 cruising at an altitude of 3 miles and never deviating from that altitude by more than 12 feet, while accurately following a flight line 16 feet wide at an unvarying speed of 123 million miles per hour.

Optical disks are removable for two reasons. First, the head "flies" more than a millimeter above the disk surface, leaving ample room to insert or eject a disk without obstruction by the head. Second, the recording material in a disk is not on the surface. Instead, it is covered by a relatively thick protective coating. As a result, the laser beam, which is focused on the recording material, is out of focus at the surface of the disk. Therefore, any dirt, dust, or scratches on the surface will also be out of focus, and will not affect the working of the laser beam. Normal amounts of foreign material, such as dust, make no difference, so the disk need not be sealed in an airtight chamber, like a fixed magnetic disk.

Beyond these basic elements common to all optical disks, each type of disk is the result of specific technical breakthroughs achieved by its developers.

Today's rewritable optical disks exploit a technology called *magneto-optics* (MO). An MO drive uses the heat of a laser beam to magnetize a tiny spot on the disk. Because it is heat that effects this change, the process is sometimes referred to as *thermo-magnetic optical recording*.

It took years of painstaking laboratory research to produce today's efficient and long-lasting thin film recording materials, which are most often alloys of terbium, iron, cobalt, and traces of other elements. The proportion of each element is adjusted to tune the medium to a certain temperature, called a *Curie temperature*, at which its magnetic coercivity drops to near zero. This simply means that it can be magnetized by a low-power, ambient magnetic field, called a *bias field*. At any temperature other than the Curie temperature, the medium is unaffected by any magnetic field, no matter how strong. The Curie temperature of an MO disk is typically around 180° Celsius (356° Fahrenheit), so under normal conditions there is no danger of accidentally erasing an MO disk, as there is with a floppy.

The head of a magneto-optic drive includes a small magnet, which supplies the bias field. The magnet's field can be switched between positive and negative, either by using a switchable electromagnet or by turning a physical magnet. When the laser beam heats a tiny spot on the disk to the Curie temperature, the spot becomes magnetized to the polarity of the bias field: positive if the field is switched positive at that moment; negative if it is negative.

Reading the magnetized spots on the disk involves a principle of physics known as the *Kerr effect*. Laser light reflected from a positively magnetized surface is detectably different from laser light reflected from a negatively magnetized surface; the angle of polarization of the light beam differs about a third of a degree between the two cases. That's not much, but it can be detected, especially if the disk has interference layers that magnify the effect to about one degree.

So much for writing and reading. Rewriting is a simple matter of returning a sector to its original polarity and then writing over it. Today's magneto-optic media have been tested through more than a million rewrites with no degradation in performance or data reliability.

### Benefits of Optical Storage

Optical disks have three characteristics that are responsible for their advantages over other kinds of mass storage: they have very large storage capacities; they are removable, providing users with access to many disks per drive; and they rotate, providing fast access to information. The large storage capacities and the removability are due to the use of a laser instead of a magnetic field to make marks on the medium. Fast access, of course, is characteristic of all disks, both magnetic and optical.

The significance of these characteristics is that an optical disk the size of a compact audio disk can store several encyclopedias with ease. One rewritable optical disk with the same diameter as a floppy diskette can store the equivalent of 650 floppy diskettes. And a 12-inch write-once optical disk can store the images of more than 150,000 typical office documents. Like optical disks, tapes have large storage capacities and are removable. But they do not rotate. To reach one piece of data, you must move past all the data that precedes it on the tape. Tape access is linear; disk access is random.

Fixed magnetic disks, such as a hard disk in a microcomputer, have large storage capacities. The largest hold up to several million bytes; but magnetic hard disks cannot be removed from the drive, and on a dollar-per-byte basis they are much more expensive than optical disks. Floppy diskettes are removable; but with a capacity of about a million bytes of data (the equivalent of about 600 typed, double-spaced pages) they don't offer very large amounts of storage space. And even though this capacity is currently being increased, it can never even begin to match that of the optical disk.

Removability provides security, interchangeability, and flexibility. At the end of the day, a disk can simply be taken out of its drive and locked up. Interchangeability means that a person can give the disk to a colleague to work on. Flexibility means that users of optical disks can share information over different computing platforms. Removable optical disks will be interchangeable among different operating systems, under standards developed by the International Standards Organization. Sony's rewritable disks, for example, can be shared easily by an Apple Macintosh, an IBM PC/AT, a Sun workstation, and a Compaq 386.

Optical disks now on the market store up to a few billion bytes, offer random access, and are removeable. This means that one could put, say, fifty 12-inch disks in a "jukebox" connected to a desktop computer and retrieve in seconds any piece of information from a database as large as 130 complete sets of the *Encyclopaedia Britannica*.

The first optical disks, which were read-only videodisks and audio CDs, became available in the early 1980s. Write-once read-many (WORM) disks have been available for almost as long. In the late 1980s, the Sony Corporation became the first vendor to ship rewritable optical disk drives in volume.

Each of these three kinds of optical disks—read-only, write-once, and rewritable—is removable, has large storage capacities, and rotates. In addition, each has its own special characteristics and benefits.

### Benefits of CD-ROM

The preeminent version of the read-only disk is the *compact disk read-only memory* (CD-ROM). Sony, along with Philips N.V. of the Netherlands, invented the CD-ROM and developed its standards.

Physically, the 12-centimeter CD-ROM is identical to the audio CD. Because it is mass-produced by injection molding, it is extremely inexpensive to make. The cost of a disk today is under $2.

The CD-ROM is an ideal medium for distributing large amounts of "static" information. The cost of distribution is in direct proportion to the value of the information, without being distorted by the price of the medium—exactly as in the case of paper, but in contrast with on-line information services (which must charge for time and connection in order to amortize large equipment costs).

CD-ROMs are physically identical to audio CDs—the same size, the same materials, the same replication process, and similar drive design. As a result, CD-ROM is far less expensive than any other medium, including paper, for distributing very large amounts of digital data.

The CD-ROMs genesis in the audio CD even determined its capacity, which is directly related to the playing time of an audio CD: 74 minutes and 44 seconds (when produced under the best possible manufacturing conditions). That duration is neither accidental nor arbitrary. It is just sufficient to record the entire length of a certain version of Beethoven's Ninth Symphony especially beloved by the chairman of Sony, Akio Morita—or so the story goes.

### Benefits of Write-Once Disks

Data, once written on a write-once disk, are permanent. A write-once disk preserves a record of all transactions recorded on it. This makes it the perfect medium for sensitive operations where it is important to keep an audit trail. All the successive versions of an engineering drawing, a contract, or a ledger page, from day to day and week to week, are permanently preserved for inspection.

In addition, the permanence of write-once disks has legal consequences. Since information on it cannot be altered, a write-once disk is more likely to be admissible as legal evidence than a rewritable medium.

Finally, the write-once disk is inexpensive. At 5 cents per megabyte, the 12-inch disk is competitive with microfilm, paper, and magnetic tape—and far less expensive than magnetic disks. Because write-once is such an inexpensive storage medium, it is very well suited for archiving information which cannot be altered and which must be stored for a long period of time.

## Benefits of Rewritable Disks

Rewritable disks take the "write once" concept and add the ability to erase and write over a segment of the disk. They are therefore more comparable to magnetic disks than are read-only or write-once disks, because information stored on them can be changed indefinitely.

A rewritable optical disk is removable, like a floppy diskette, but instead of holding about one megabyte (like a floppy) or even 30 megabytes (like a typical microcomputer fixed disk), one double-sided rewritable optical disk holds 325 megabytes per side. A single 5¼-inch rewritable optical disk holds more information than twenty 30-megabyte hard disks.

There are many untapped uses for this kind of storage medium—much as when hard-disk drives were introduced. When the first 5-megabyte drives came on the market, in 1982, few people could imagine filling up that much storage space with the personal computers then in use. But the limitations of those PCs were due in part to the unavailability of mass storage. As soon as such extensive storage became available, software programs began to take advantage of these capacities. The PC user soon had access to software applications formerly available only on minicomputers and mainframes. Database management systems, powerful spreadsheets, graphics programs, and other software that produced large files became feasible. In those days, a quarter of a megabyte was a very large file on a PC.

Today, there are plenty of specialized applications—especially those generating graphic output—that produce very large files. In addition, as PC users gain proficiency in creating files, there is an increasing tendency to retain them. Users are very reluctant to delete anything—even files they will surely never use again. Rewritable optical disks, with their enormous capacities, will foster the development of new applications that generate even larger files (such as video applications, which are already on the horizon). That is the nature of the computer revolution.

What are some of these present and future storage-intensive applications? They are numerous and varied, and they all require one or more of the characteristics of rewritable optical disks: capacity. longevity, lack of susceptibility to damage, and (perhaps most significant) removability.

### Multimedia (Voice, Video, and Data) Applications

There are over 1000 different commercial CD-ROM disk titles currently published worldwide, but the most exciting CD-ROM products are just coming to market. They employ a technique that will eventually make interaction between humans and computers more intimate than ever: *multimedia,* the synergistic use of text, voice, music, video, graphics, and other kinds of data, rather than just text alone. Using all these types of data together, the computer will be able to present information with a richness approaching that of human-to-human interaction. Only optical storage has enough capacity to store and deliver multimedia.

To understand the concept of multimedia, imagine reading an encyclopedia entry about John F. Kennedy. Merely reading the entry is a pale experience when compared against reading the entry and then seeing a photograph of Kennedy while listening to a recording of one of his stirring speeches. This kind of experience can be achieved only by means of multimedia delivery, in which various forms of information, heretofore requiring separate and distinct media, are presented on the same medium. In fact, a prototype of an entire encyclopedia on CD-ROM applies just this multimedia technique to articles on a variety of subjects, including John F. Kennedy and the National Football League.

One feature of this encyclopedia is something called *hypertext.* Hypertext requires large amounts of data to be retrieved very quickly and is a good example of a new computer application made possible by optical storage. Consider the example of the encyclopedia article on John F. Kennedy. The term "Democratic Party" might occur within the article, highlighted in green. By moving the cursor to the term and selecting it by pressing the Enter key or clicking the mouse, the user could cause the Kennedy article to be replaced on the screen by an article about the Democratic Party. When the user is finished with that article, a simple key press or click returns the Kennedy article to the screen. These hypertext routes from one piece of information to an associated piece can be built into the CD-ROM carrying the information. Moreover, with writable media the user can create additional pathways through the existing information, as well as pathways to his own annotations, customizing a body of information to his own precise needs.

This kind of feature would save untold hours of research. For example, a financial analyst could create links among various databases describing different aspects of a company. It would then take only moments to move from the balance sheet of Acme Widgets, as stated in the firm's annual report, to an item in a year-old newspaper account describing the firm's involvement in litigation with the Securities and Exchange Commission.

Another prototype CD-ROM contains an educational section on the solar system in which is displayed an artist's rendering of the sun and its surrounding planets. The user is asked by a recorded instructor to select any planet. When the user selects Saturn, the screen displays a series of satellite photos of Saturn and of its rings and its moons, as the instructor explains some of the exotic circumstances of this planet. Celestial music plays in the background. This kind of application is based upon the well-known educational rule that the more ways in which information is presented, the better it is received, understood, and remembered.

Multimedia is very new to the computer world, because it requires more storage capacity than could be delivered before optical storage. Of all the different kinds of information, text as it is typed into a computer takes up the least storage space. After text, line drawings are the least voracious users of storage space. Sound and images gobble up storage capacity beyond the ability of magnetic storage to provide it. Until recently, no storage medium has had the capacity to hold different kinds of data economically. Beyond the data, hypertext links take up storage capacity and, in fact, present a storage requirement that never before existed.

The storage capacity of CD-ROM, or any storage medium, can be augmented by compressing data, which allows much higher effective storage densities. Development of international standards for compression and decompression is underway. Once these standards are set, multimedia will become a viable new industry for entertainment, educational, and business applications for the home and the office. Such products will take advantage of the merging of television and computer technologies, and of advances in digital phone service, to deliver computer-controlled multimedia programming to new markets.

But most profoundly, computers that are capable of multimedia will create a different psychological relationship with their users. Computers will more closely resemble people. Clever programmers, employing the centuries-old skills of novelists and actors, may become adept at using techniques such as the creation of individual personalities for computer programs to break down the barriers between man and machine, reducing the intimidation factor and encouraging people to interact with computers in a relaxed manner.

Few multimedia products have yet been commercialized, but existing CD-ROM applications are nonetheless exciting in their own right. Many are large databases of text and images. Some are business databases with extensive profiles of all the corporations traded on the New York Stock Exchange, including recent financial figures, names of officers, contact information, descriptions of products and services, and images of documents such as annual reports and 10-K statements.

Another successful type of CD-ROM product is a parts catalog. In the automotive field, such systems are now used by independent mechanics and by dealerships for General Motors, Ford, Honda, Chrysler, Isuzu, and other carmakers.

Until the advent of CD-ROM, the parts clerk at a dealership looked up ordering information in thick, cumbersome, grease-stained, dog-eared paper catalogs or on faded, scratched microfiches. Today, the entire parts catalog for a car model can easily fit on a single CD-ROM, along with indices, lists of parts numbers, automated ordering information, and diagrams at several levels of magnification.

After information of this kind became available in machine-readable form, thanks to CD-ROM, it was possible to integrate it with electronic systems for ordering, inventory control, billing, and bookkeeping. Dealerships that have installed CD-ROM parts systems report impressive savings: the right part gets ordered more often and in less time. CD-ROM parts catalogs have not only speeded up parts handling. They are initiating changes within the car dealership itself, the most visible of which is that wholesaling has grown in importance among dealers using CD-ROM.

Many dealers not only use parts to service cars, they also sell those parts on a wholesale basis to other dealers and to independent mechanics. Dealers using CD-ROM parts catalogs (many of which use disks pressed by a Sony subsidiary, Digital

Audio Disk Corporation) can take orders from their wholesale customers in about a third of the time it took when they used paper catalogs. With the time they save, they are expanding their wholesale business.

Another way in which CD-ROMs are changing the organization of car dealerships is by enhancing the roles of the mechanic and the service manager, who now find it much easier to order parts themselves rather than through a parts manager. Parts managers, naturally, see this shift as threatening. This is a modest reminder that implementing technological changes in the workplace will also involve sensitive institutional considerations.

Multimedia technology will bring even greater benefits in the field of automotive service and repair. A CD-ROM will contain not only all the parts information but also video footage of proper repair and maintenance procedures and voice recordings of master mechanics describing the tricks of their trade. The use of video and voice is especially important where users may be illiterate or may not speak English. A CD-ROM can contain versions of the same audio information in multiple languages, and the user can select the one he wants. There are further uses of multimedia delivery for the blind, the hard of hearing, and other handicapped workers.

The automotive service field provides only one of many examples of ways in which computers encourage businesses to be more productive. Computers are having the same effect on business that tractors had on farming. The trend is for the average business to grow in size, just as the average size of farms has been growing for decades. Businesses such as retail merchandising, which were most sensitive to the efficiency of automated transaction processing, have exhibited this trend for some time. Wal-Mart and Toys R Us are well-known examples.

In a few years, a variety of organizations using parts catalogs will take advantage of the benefits of CD-ROM. Large retailers will be freed from the hard-to-read microfiche. Their customers will select products from crisp, clear color images displayed from CD-ROM. Fewer parts-desk clerks will be required to look up functions. They can instead engage in sales, which will be facilitated by the appealing images the customer is browsing.

Multimedia CD-ROMs will also be used for point-of-sale applications. To a limited extent, they already are. For example, a product for travel agencies lets a customer browse through still, color images of various resorts and select information about pricing, accommodations, travel schedules, local customs, and points of interest, all stored on a CD-ROM. Hotels, restaurants, airlines, car-rental agencies, and other vendors are not about to overlook the possibilities for advertising in such a medium.

CD-ROMs will also be used for automotive navigation. The top-of-the-line Toyota Crown, sold only in Japan, already offers an optional navigation device featuring a CD-ROM and a small computer monitor in the dashboard. A digital map is displayed on the monitor. A cursor indicates the car's position. The push of a button replaces the map with various kinds of information about a given geographical area, such as where to find a gas station. In fact, Japanese government agencies are working with a consortium of corporations to develop a nationwide system of sensors. This system might be built into roadways to provide more accurate information, through the on-board navigation system and other automotive computers, than is now being supplied by either satellite or cellular technology.

Similar CD-ROM navigation systems are being developed in the United States and are, in fact, already available for maritime use. The greatest benefit will be to large transportation companies which must manage fleets of trucks or ships. Future navigation systems will be based on CD-ROM maps linked to satellites. Improved navigational precision for ocean-going vessels will yield significant savings in fuel costs. Trucking companies will enjoy savings in both time and fuel from better and more flexible routing. A small computer in a truck could use CD-ROMs to store detailed maps of the entire country, which could be augmented by satellite-transmitted updates on detours, weather conditions, and so forth.

Another important use of CD-ROM is already becoming established. Several major computer companies (notably Sun Microsystems, Apple Computers, and Hewlett-Packard) use CD-ROM to distribute operating-system software for their large computers, including complete technical documentation and training information. In this application, CD-ROMs are paying for themselves just by the paper-storage, handling, and production costs they are saving.

Since technical documentation is distributed on the same CD-ROM disk as system software, instead of in a shelf-full of manuals, a systems manager has everything in electronic form, available for automated search and immediate retrieval. Since factories are now offering one-day turnaround on CD-ROM processing, a manufacturer can change its software and technical information as often as it wishes. Responses to system problems will be faster, reducing downtime significantly and leading to organization-wide increases in productivity.

Use of CD-ROM for software distribution is also just beginning to spread to retail operations. One firm is developing a system to distribute popular applications, such as spreadsheets, word processors, and database-management systems, on CD-ROM. A 600-megabyte CD-ROM can store hundreds of even the largest software packages.

A store distributing software from CD-ROM would be much smaller and cheaper to run than current retail outlets. Instead of buying software prepackaged on floppy diskettes, the customer would simply copy it (along with complete documentation) from the CD-ROM. Instead of paying overhead on a store full of manuals, the software store of the future may stock only CD-ROMs.

And more mundane uses should not be overlooked. For example, Xiphias, a developer of CD-ROM applications, has already produced a CD-ROM containing the white pages of all US telephone directories, and the northeastern telephone utility NYNEX has released a CD-ROM containing the white pages of New England and New York. Reports of early demonstrations note that when most people use the national white pages for the first time, they look up former romantic interests with whom they have lost contact.

### Write-Once Applications

As versatile and useful as CD-ROM is, we are accustomed to disks and tapes on which we can write information ourselves. The 12-inch write-once optical disk has begun a revolution in the storage and retrieval of all kinds of documents. Sony's High Capacity write-once system, for example, stores 6.55 GB (equal to 6550 floppies) on a single disk at the very low cost of 5 cents per megabyte.

Write-once disks record data permanently by making physical marks on the disk. Some drives turn tiny holes in the disk; some create tiny bubbles. The most reliable and longest-lasting recording mechanism for write-once melts two different layers so that they fuse into an alloy. This tiny spot of alloy reflects a laser beam differently than the unfused material around it, permitting the write-once drive to distinguish marks from their background and convert the pattern in which they occur on the disk, which is a kind of code, back into data. Sony is the only company using this recording mechanism for its write-once drives, which is why Sony guarantees its write-once disks to last 100 years.

The most important market for write-once disks is electronic imaging, which permits the automated storage and retrieval of documents. Endless rows of filing cabinets and microfilm bins will be replaced by optical disks in "jukeboxes," which will take up far less space.

The printed page is an analog medium, a picture of speech. Each letter on the page is the analog representation of a sound; each written word is the analog representation of an element of spoken information. In some cases the analog nature of letters is more obvious than others—for example, the letter O has the same shape as a mouth uttering the sound it stands for. The page is an image. When the page is transferred to microfilm, the image is reduced by photography—a process designed for use with images.

Computers, on the other hand, have until recently handled information as numbers. Text has been stored and processed in the form of numeric code in which each character on a keyboard is assigned a unique number. In the most common text code for personal computers, the American Standard Code for Information Interchange (ASCII), a capital A is 65, a capital B is 66, and so on. Computers can also handle information as images, but that requires more storage space. Until optical disks were developed, many users considered that space to be too expensive.

Handling information as code would be perfectly adequate if all text were created on a computer, as it someday may be. At present, 95 percent of the world's information, even that part created by computer, is stored and transmitted on paper (including invoices, bank checks, letters of credit, shipping waybills, contracts, remittances, fund transfers, and engineering drawings). All this paper occupies enormous amounts of real estate. This is yet another cost of current storage procedures.

Perhaps the greatest cost of analog documents is the inefficiency of retrieving them. The state of the art of document storage and retrieval for most businesses is still the manila file folder or the hanging folder. The most conservative analyses show that 3–5 percent of all documents in folders are irretrievably misfiled at any given moment, and that it costs an average of $250 to rebuild a document. Simply owning and operating a filing cabinet can cost up to $685 a year (Gerry Walter, "Optical Storage of Office and Transaction Documents," Rothchild Consultants, 1988). This too may be a conservative estimate.

Even harder to assess is the cost of the productivity lost when a file is "out." Several years ago, the public relations department of the Ford Motor Company was in the process of cost-justifying a write-once-based system to store its archives of speeches, news clippings, and other publicity-related documents. The department was so immersed in paper that the department manager was compiling the costs of storage per document, the man-hours spent retrieving the average document, and the costs of paper versus optical disks and other variables. Preliminary findings indicated that an automated storage and retrieval system based on write-once disks would save a lot of money. But management has not yet made its decision, and the switch to optical disks was far from certain. One day, an executive came to visit the department in quest of a copy of a speech. While searching through file drawers, closets, and stacks of cardboard boxes, she opened a two-door standing cabinet. A box full of documents fell off the top shelf onto her head. She wasn't seriously hurt, and the speech was eventually found. The optical disk system was approved the next day, and the department has enjoyed immense productivity gains ever since—to say nothing of a considerable reduction in stress.

Electronic imaging combines several technologies—the key one being optical storage—to store all information not on paper but on write-once disks, and to retrieve it automatically. The first step is to digitize the page, preserving its pictorial nature while converting it into a form the computer can handle. A scanner takes an electronic picture of the page. This picture consists of tiny dots where the page is black, and nothing where it is white. The number of dots per inch may vary, but the most common figure, more or less standard for document imaging, is 300 dots per inch. That means that in a square inch that was all black there would be 90,000 dots. Typically, each dot is represented by a bit of data, which equates to 90,000 bits per square inch, or one megabyte per 8½ × 11-inch page.

Of course, most of a typical page is white, and with compression techniques (originally developed for facsimile transmission) the average page of an office document takes up about 50,000 bytes—still a hefty chunk of storage in a corporation where hundreds of millions of filed documents are active. The statistics in such a case are formidable. A four-drawer file cabinet stores, on the average, 1300 pages per square foot of office floor. Including the required operator space, it consumes 7.7 square feet of floor space. Mechanized files—machines that allow either storage up to the ceiling (bringing the shelf down to the user as needed) or movable storage shelf units (eliminating the aisle space between the shelves)—can be up to 10 times more efficient. Such a four-drawer file cabinet holds about 10,000 pages.

A key technology working in tandem with write-once is Workflow software. Workflow, an office automation tool, is software that automatically routes a document from desk to desk, tracking its progress in terms of schedule, supervisorial review, and so on.

Consider an insurance claim form. Today, it is received by mail, taken out of the envelope, stamped with the date, assigned to a supervisor who assigns it to a claims adjustor, put into an account file, evaluated, passed on to the accounting office, and so forth.

In a growing number of offices, as soon as a document is received it is scanned, digitized, and stored on a write-once disk. This disk is then placed in a write-once "jukebox," and the file can be retrieved at will by any computer on the network. How much storage does this afford the user? Sony's jukebox holds fifty 12-inch write-once disks, providing 327 GB of on-line storage. And up to seven of these jukeboxes can be "daisy-chained," giving the user access to 2.2 TB of information. (1 TB, or terabyte, equals 1000 GB; thus, 2.2 TB represents 2,200,000 floppies.) From that point on, the document is handled automatically by workflow software. When a supervisor turns on his or her terminal in the morning, workflow software automatically displays these electronic documents, in sequence, for assignment at the press of a key.

When the worker turns on his terminal, the documents assigned to him are displayed one by one for disposition. Thus a document passes through the organization, never missing a step, staying on schedule, never becoming lost or misfiled. When a customer calls with an inquiry, the document can be displayed immediately by account number, customer name, date, or whatever information has been used in indexing it. The time it takes to satisfy customers' inquiries in workflow environments using write-once systems has been cut from days to minutes, and the number of phone calls has seen a similar reduction.

Because of their low cost, high capacity, and longevity, 12-inch write-once disks are the foundation of document-storage systems. But text documents are only the beginning. Equally important are other forms of information that must never be altered and need to be stored in vast amounts for long periods of time.

Until 12-inch write-once disks made it possible to automate this information, it simply went unused. Consider the data from satellites, which constantly beam down telemetry-borne data in streams so wide and fast that it often is impossible to capture them all. Until the advent of large-format write-once disks, not only was some satellite information uncapturable, but much of it was unusable—not because it was degraded, but because there was so much of it. Information was so expensive to disseminate that researchers simply could not find the data they needed (or if they could, they could not easily or economically gain access to them). A scientist in Berkeley often had no reliable way of learning what data were gathered by an institute in France, a government agency in Japan, or even a university on the other side of San Francisco Bay.

Another example is provided by the vast amounts of seismology data that oil exploration companies have gathered over the decades. A major oil company typically has in storage about 750,000 reels of tape in one department, and that is only a fraction of the company's total seismic database. Each tape must be rewound and copied periodically, at great cost. It is extremely difficult to keep track of the information on all those tapes. Even when you know which tape you want, it can be cumbersome to locate it. Using the data is, at best, inconvenient and expensive. But 200–300 reels of tape can be transferred to a single 12-inch write-once disk. In every case where this has been done, the result has been a leap in the use of data and a resulting increase in available oil. For example, oil companies share geophysical information about adjacent territories, each using the other's data about the terrain to increase the efficiency of its own wells.

Both satellite-generated data and seismic data are unique, timeless in value, expensive to capture, and crucial to certain business operations, whose owners cannot afford to have data accidentally (or deliberately) modified, erased by a stray magnetic field, or lost when a computer system "crashes."

The timelessness of data is often overlooked. As more efficient methods are developed, the incidental data of 20 years ago become the priceless information of today. The value of all those oil company tapes has increased along with the price of oil and the value of unexploited reserves. The value of old information can also increase when a new technology comes along in a realm other than computing. For example, a company may have data collected 30 years ago on an oil field that has since ceased to produce efficiently. Such data may not appear to be of any further value. But then along comes a new method of tertiary oil retrieval involving injected polymers, accompanied by sophisticated graphics-based simulation methods by which a scientist can model the effects of various polymers under various pressures to get the optimum combination for retrieving residual oil from that field. Suddenly the old data become very valuable. Moreover, optical disks turn out to be the best storage medium for the enormous amount of data created by simulation software. New data are created by combining old seismic data with new information about the underground behavior of polymers in reservoirs.

Similarly, the value of information can be proportional to the number of people who have access to it. A medical database may lead to a life-saving drug if researchers around the world can collaborate on it. But if the information is so difficult or expensive to distribute that only a dozen researchers in one location can use it, that drug might never be developed. Imagine the power of a conference of medical researchers where each attendee carried with him a disk or two containing hundreds of megabytes of information, which could all be exchanged with his colleagues through small, inexpensive computers in a special room at the conference. As information becomes more accessible, the market for it will grow, leading in turn to the gathering of even more information and the discovery of more knowledge.

Until write-once disks became available, the immense information bases of seismic data, satellite data, and similar kinds of information rested in endless racks of computer tape, retrieved, if at all, by entry-level workers, manually, in a system often referred to as "sneaker net." Write-once-based electronic imaging systems will allow organizations using "sneaker net" systems to automate un-cross-referenced, document-based image data and distribute it electronically.

Write-once disks, like all optical disks, are removable. Since data on a write-once disk are also unalterable, it is the preferred medium for exchanging data that must be intact. Consider the US Patent and Trademark Office (PTO), which uses a large write-once system to store documents and to speed up patent approvals. Write-once will also allow the PTO to exchange patents with European and Japanese patent offices almost instantaneously, perhaps foreshadowing a single global patent office. The US Federal Drug Agency, to give another example, is accepting new-drug submissions from pharmaceutical companies on write-once disks. Under this system one disk arrives at the FDA, rather than a truck carrying a ton of paper detailing the company's research and testing on a new drug. Not only is the medium more compact and guaranteed to be more accurate, but the disk includes a cross-reference index matrix, prepared by the pharmaceutical company, for automatic search of the documents. As a result, the submission can be evaluated quickly, with incalculable benefits for patients.

Images are playing an increasingly important role in other areas of medicine. X-rays and digital images from magnetic resonance, ultrasound, and computer-aided tomography (CAT) equipment are expensive to store and difficult to retrieve. Moreover, they often tie up, in the form of photographic film, silver that would be worth hundreds of thousands of dollars to revenue-starved hospitals (at no expense to patients). Electronic image management provides a way to recover this expensive but overlooked asset.

A busy radiology department can easily fill a bank of large hard disks in a single morning with the output from digital-modality (e.g., CAT and MRI) equipment. The data have to be transferred to tape before being analyzed. This reduces equipment use and delays diagnosis. On tape, the data are often too cumbersome to retrieve months or years later, so the physician is deprived of potentially valuable insights into a patient's history.

Write-once medical systems are already used by many hospitals. They not only solve storage and retrieval problems but also permit the fast exchange of images throughout a facility. Manufacturers of CAT and MRI equipment and other digital medical tools are about to introduce models with built-in ports for write-once drives, which will make the integration of systems faster, easier, and cheaper.

In the future, hospitals will be able to collect data on all patients in real time, via monitoring equipment and connections at each bed. Data will constantly be analyzed, and will also be stored in write-once jukeboxes for future reference. Images can be pulled off disks in a jukebox quickly and sent to another department or another hospital halfway around the world. Subsets of a patient's records, including images, will be downloaded to smaller media (probably rewritable optical disks) for distribution to personal physicians or to the patient himself.

Another health benefit concerns the storage of long-term data. Data on hazardous waste sites, for example, may have to remain readable for centuries, and they must not be altered.

### Rewritable Applications

One of the most important computer trends is the trend toward local-area networks connecting all the computers and peripherals in an organization. LANs represent the achievement of several ideals: an increase in the sharing of electronic information among co-workers, a breaking down of the barriers of incompatibility between different kinds and sizes of computers, and the emergence of a completely new work environment. As one computer-industry CEO has said: "The user will be able to access data without knowing—and, more important, without caring—where that data is stored. The PC will not act as simply an extension of the mini or mainframe. The mini or mainframe will be an extension of the PC."

LANs will increasingly supplant mainframes. This has extraordinary implications for the use of rewritable optical storage. For example, a LAN connecting twenty PCs using Intel

80386 microprocessors provides more combined processing power, as measured in MIPS, than most existing mainframes.

MIPS aside, LANs do not yet offer throughput to match the channel speeds of mainframes, but the day may come when they do, thanks largely to fiber-optic connections. When that day comes, LANs may well offer processing power, data-transfer rates, and (thanks to retrievable optical storage) mass storage capacity rivaling mainframe computers at a fraction of the cost. An added benefit would be flexibility: when a mainframe goes down, the whole system goes down; when a LAN workstation goes down, the other workstations keep on running.

Besides bringing the storage capacity of mainframes to LANs, rewritable optical disks will bring interchangeability. Because a worldwide standard is being developed by the International Standards Organization for rewritable optical disks, the convenience of rewritable optical storage will eventually extend to platform interchangeability. All rewritable optical disks will be interchangeable among different operating systems. Recently available software provides disk interchangeability even without further improvements in the ISO standard. This software translates among the file structures of various operating systems, making it easy to read and write data on the same disk, with, say, an IBM PC, a Sun workstation, a DEC MicroVAX, and an Apple MacIntosh. As optical storage's capacity can help turn LANs into substitutes for mainframes, their interchangeability will help transfer information among different LANs.

Connecting people within networks and connecting networks within bigger networks multiplies the sharing of information. Shared information begets more information. When two pieces of information are combined, the result is often new information. For example, if you have the unit orders for your company's products for the past 24 months and someone else has the currency fluctuations for the same period, together you can create a graph showing the correlation, a report analyzing the implications, and a memo suggesting changes in the company's sales operations and in its asset-management policies.

The advantages of security, interchangeability, and flexibility will be available not only at individual workstations. Rewritable optical disks will be used to hold all the data used on a network in a single central repository. Here, removability will eliminate the headaches of backing up data. The file server will contain two disk drives. All data will be *mirrored*—that is, written identically to both of them. At the end of the day, one can then be taken out for backup storage and the other left in for the next day's work and changes.

Finally, LANs and wide-area networks (WANs) will become more efficient. Moreover, the global telephone network will become a digital network to which all other networks will connect. Interconnection and integration of computers with other digital devices, such as video displays, telephones, and fax machines, will occur at the same time as network interfaces become standardized, fiber-optic cable increases the amount of information that can be transmitted, and more powerful processors enhance and localize control over network functions.

Our ability to share information will leap, but it will also demand some way for each of us to store personal information. That way will be rewritable optical disks. In fact, without rewritable optical disks the full potential of the coming network age might not be achieved.

Rewritable optical storage will also permit new applications where sharing information is less important but where the ability to alter data (especially images) is required. Some of these possibilities already exist. Suppose a landscape architect is designing the grounds of a new house in the mid-1990s. Data on the terrain have already been gathered and stored on large write-once disks, with subsets distributed on CD-ROM for use by government planning agencies, by the utilities responsible for installing the water and sewer pipes, the gas lines, the electrical cables, the fiber-optic multiple-data telecommunications cabling, and by the contractor who will grade the property. The landscaper buys a subset of the data on a small-format rewritable disk for his PC. Using simulation software and a gardening database on a CD-ROM disk, he instantly generates views of the future of the grounds. He simulates planting a maple tree at one location and views an image of what it will look like and where its shadow will fall in five years. He sees that it will shade the vegetable garden, so he moves it to another corner of the grounds. He selects plants to fit the terrain and the climate, and he moves them around on screen until he is satisfied. Each time he makes a move he rewrites a portion of the evolving plan, making good use of the rewritable capability of his optical drive. When he is done, he orders seeds, bulbs, and saplings electronically from a list that his software compiled on the rewritable disk as he worked, based on pricing contained in the original CD-ROM gardening database, which also contained advertising for various nurseries that offer order-by-modem services.

Today there are image-simulation applications in use that require optical storage's capacity. For example, a weight-loss clinic can take a patient's photo, use a scanner to digitize it, store it as a computer file, and then use image-editing software to remove excess pounds, to add makeup, to change hairstyles, and perhaps to dress the patient in a new wardrobe. The result is before-and-after photos at the beginning of the therapy instead of at the end. And this can be a powerful motivator for the patient's health.

Such imaging systems use a lot of storage. Because the images need to be altered, rewritable optical is the perfect medium. The use in plastic surgery is obvious. Applications are also being used by law-enforcement agencies to automate the creation of "artists' renderings" of suspects. They are used by advertising agencies to test-market designs for ads and labels. They are used by film and video production companies to create story boards and to evaluate designs for sets, costumes, and lighting.

Image synthesis and design applications will have other uses. Business presentations will easily be customized to each audience. An architect will try out different versions of a building on a screen. A mechanical engineer will play with the design of a product, its subassemblies, and the parts of each subassembly until he achieves the balance required by his manufacturing and financial constraints.

There are other applications for rewritable optical disks that have nothing to do with images. Hotels keep records of their customers. A database of all the customers who stay in a large hotel over the course of 10 years, for example, is too big to manage economically on magnetic disks. Rewritable optical disks will probably be used in such systems, and tied in with the hotel's reservation-management system. In some cases retrieval of the guest's record will be automatic, in response to the phone number from which he calls, which will be delivered to the called party as a feature of telephone service. This kind of service is already available in selected locations and will be used nationwide by the early 1990s. Such precision helps th hotel to remain competitive. The same holds true for hospitals, both for cost cutting and for physician referrals. Consider the possibilities of applying the same techniques to telemarketing, shopping by phone, or any other aspects of customer relations.

Rewritable optical storage is already inexpensive, and is getting more so. Comparing optical and magnetic rewritable hard disks is tricky, because magnetic disks are not removable—to get the disk, one has to buy the drive. Since optical disks are removable, the user needs only one drive to use any number of disks. The comparison on a dollar-per-megabyte basis is dramatic. A 650-megabyte rewritable optical disk costs about $250, or $.38 per megabyte. A rewritable magnetic disk and drive with a comparable capacity would cost about $4500, or $6.92 per megabyte.

The price of rewritable optical technology will fall rapidly. A few years ago, a videocassette recorder cost $1500. Today, comparable models cost a tenth of that. Semiconductor memory prices drop by half every 2 years. Rewritable optical will enjoy a similar cost curve, and will eventually become so inexpensive that it will be a standard peripheral, as a floppy-disk drive is now. In a very few years, we will be using laptop computers with built-in rewritable optical drives and 3 inch disks, carrying in our pockets as much information as the National Archives now store in several rooms.

### Combined Applications

Perhaps the most intriguing applications of optical storage involve combinations. In a few years, mass-market microcomputers may have only two forms of storage, and both will be optical: a rewritable drive and a CD-ROM or some similar read-only drive. This combination will be adequate for all the needs of a general-purpose business workstation. In such a workstation, mass storage performs three functions: data distribution, working storage, and backup. Data distribution will be shared by CD-ROM and rewritable optical disks. CD-ROM will deliver reference databases and very large software and documentation packages. Rewritable optical will deliver small software packages, as well as working storage and backup.

Farther down the road, all these functions will be combined into a single small-format drive capable of using both rewritable and read-only media. For example, within a few years we might see a drive that can read, write, and erase rewritable disks that are about 3½ inches in diameter and can read CD-ROMs of the same size.

A multi-function or hybrid drive using both rewritable and read-only disks opens the prospect of future computers using optical disks as their only form of mass storage. Such a drive, using disks 80–90 mm in diameter, would be faster, since the head would have less disk geography to cover and because the lighter disks could be spun much faster. Also, technical advances will eliminate the need to erase and rewrite data on a magneto-optics (MO) disk on separate revolutions. The performance of such drives will match or exceed that of magnetic disks. They will have greater capacities and will cost only slightly more. We may well see a single disk containing a section of read-only data and a blank section that is rewritable. Such a hybrid MO/ROM disk would be made of polycarbonate plastic, just as CD-ROMs and MO disks are now.

A hybrid disk would be the ideal medium for distributing large databases meant to be used differently by each user. Recalling our discussion of hypertext, imagine one disk containing a read-only version of a whole bookshelf of reference material and a rewritable workspace on which a person could add his own annotations, pathways, and additional data files. No magnetic disk could ever provide this kind of flexibility, because data distributed on magnetic disks must be written to them electronically. Even on high-speed copying machines, it is impossible to write data to magnetic disks at anywhere near the speeds with which hybrid MO/ROM disks could be stamped out.

Offering unique advantages, a hybrid optical disk would not have to compete with magnetic disks on performance. Such a disk, used for data distribution, would have to adhere to a recognized standard for compatibility and interchangeability. The International Standards Organization has developed such a standard for MO disks, and disks from Sony and 3M support that standard (as with others as they come to market). A similar international standard will no doubt be developed for hybrid disks.

Besides disks, the principles of optical storage can be applied to strips of flexible media. In lengths of hundreds of feet, these will be used on reels as optical tape. In lengths of 2 or 3 inches, they will be affixed to pieces of plastic, like credit cards.

Optical tape will be used to capture the literally astronomical amounts of data sent back from satellites. One of the first such facilities was installed at the Canadian Center for Remote Sensing in mid-1989. Optical tape may also be used to record high-definition television, and for a myriad of other uses demanding huge capacity without random retrieval. At the other end of the spectrum, a single optical card, which will be carried in one's wallet, purse, or pocket, will store about 700 typed pages or about 30 newspaper-quality photographs, or a combination of the two.

Optical tape is barely on the verge of commercial use, and optical cards are somewhat farther away. Optical disks, however, are more emphatically on the market today.

### Conclusion

Optical storage, like any other technology, is most significant in the ways it enhances, complements, and supports related technologies. That is just another way of saying that technology is important not for its own stake but as a means of satisfying human needs and accomplishing human goals.

In that light, optical storage supports a historical trend of the past few decades—the trend toward digitizing information to make it easier to automate its transmission and processing, thereby making it available to much wider groups of people. The full power of computers can be applied only to digital information, and for many years the world's information has been undergoing a quiet transformation from the analog state to the digital. In the future, virtually all information will originate in digital form, whether it is video, audio, x-rays, satellite information, or graphics. All information will be accessible to computers—that is, to any personal workstation with sufficient mass storage.

Since information will be created in digital form, it will be accessible in its original condition. Slides created at a personal workstation using desktop presentation software will look exactly the same to their eventual audience as they do at the moment of their creation. A page of text, typeset in any combination of fonts, containing any sort of special graphic effects, will look exactly the same on the printed page as on the screen where it was created. Until now, we have automated communication only by paying a penalty in design constraints.

These new capabilities will vastly enhance the effectiveness of our electronic communications. They cannot happen without mass storage. For example, a powerful desktop publishing system based on digital typography and graphics requires all three kinds of mass storage: read-only, rewritable, and write-once. CD-ROMs will be used as on-line databases holding enormous files containing limitless digital typefaces and fonts, catalogs of clip art, dictionaries, thesauruses, books of quotations, and design handbooks. Rewritable disks will be used to store works in progress, from single pages to large volumes, including multiple versions pending choice of a final design. Finally, when the work is in its final form, it will be written to a write-once disk for storage and reference. In addition, wrote-once disks will be used to build in-house libraries of customized alterations of materials distributed originally on CD-ROM.

Because there is so much information in the world, and because it is so hard to correlate, the only way to use it meaningfully is to specialize. Yet specialization has its penalties. It tends to reduce communication between people working in different disciplines. At a time when solutions to global problems seem to demand cross-disciplinary cooperation and the sharing of ideas, it is vital to balance specialization with enhanced communication, to make more information accessible to more people in more ways. Toward this end, optical storage will justify the billions of dollars that have been invested in it.

# 9

# *Imaging Capabilities in the 21st Century*
*Harold F. Langworthy*

Solemn prophecy . . . is obviously a futile proceeding, except in so far as it makes our descendants laugh.
*J. B. Priestly, 1927*

When Neil Armstrong stepped onto the surface of the moon, most of the world could watch via live or delayed broadcast. Even those who were not able to understand his words could share the experience. The Apollo missions returned magnificent color photographs which allowed us to see the moon's surface as if we were there, and to see the earth rising over the moon's horizon. These images made the missions real for us, and no amount of text or commentary could have replaced them.

Images are so important to communication that media using only voice or text have been transformed when the technology needed for imaging has become available. Broadcasting began with radio, but television is now our dominant form of in-home news and entertainment. Gutenberg's press made it fast and inexpensive to print words, but pictures required so much more effort and expense that until recently newspapers, magazines, and books consisted almost entirely of text. Today, however, most newspapers compete for the latest, most dramatic photographs, and some newspapers, magazines, and books consist primarily of photos accompanied by a small amount of text. Even scholarly journals often have ads full of pictures.

During the 1990s, advances on technology will enable us to integrate images into computer-based information systems, producing dramatic enhancements in the usability and the power of these systems. In the longer term, we will develop integrated information environments that will make full use of text, graphics, images, and voice. These integrated environments have the potential to transform the workplace, resulting in a richer, more productive, more interesting environment.

Images serve us in many ways, and each of these will add value to computer-based information systems. The sayings "Seeing is believing," "See for yourself," "I saw it with my own two eyes," and "I'm from Missouri—show me!" demonstrate the credibility which images possess, and this credibility will persist even if the images have been digitized, processed, and reconstructed. Images are also essential to imagination. Besides continuing to be important in entertainment, they will allow us to see the results of complex calculations and to visualize the answers to "what if?" questions in an imagined world.

Images are also vital to preserving information. Many of our most vivid memories are visual, and much of human history has been preserved in drawings, paintings, and photographs. Oral history can be as fragile as human memory, and written history may not outlive its language, but the cave paintings at Lascaux and Altamira speak to us over a gulf of tens of thousands of years. The tombs of the pharaohs or the Etruscans, the buried army of Qin Shi Huangdi, and the reconstructed village of Pompeii provide vivid and valuable information, largely because of the images they contain. Information systems will be required to handle these images, not just to describe them with words.

Images are also an important medium of communication, conveying large amounts of information quickly. We say "A picture is worth a thousand words," and we ask "Do you see what I mean?" Signs with picture symbols are planted alongside our highways to advise travelers of road conditions in a universally understood language. Advertisements usually contain some text, but the real message is often in the image. Even mathematicians and physicists who work with abstract concepts use visually oriented terms, such as "black holes" and "strings," to describe their ideas.

The impact of imaging will not be limited to the use of images as a particularly rich form of information. Early computers used a command-line interface, and every user had to learn the syntax of the operating system in order to carry out even the simplest function. A cryptic "C>" provided no clue about what to do next. More recently computer systems have used pictures, or icons, that represent objects or functions. The desktop computer uses images to draw an analogy between what it can do and what we want it to do. We understand the functions of a file folder and a trash can; thus, we readily grasp that if we put a file folder into a trash can the information in the file will be discarded.

The use of images in guiding our access to computers will become even more important as the functions performed by computers become more complex. The desktop analogy may expand to include a whole office or company. Complex functions such as manufacturing and distribution may be planned and scheduled using images, just as models have been used in war games by the military. Imaging will not be a separate, add-on function. It will form an integral part of a more natural, more useful, and more interesting information environment in which computer use will no longer be an acquired skill.

In this new environment, will seeing still be believing?

We can now digitally process motion-picture film to enhance the images. The first applications have been the generation of special effects for films like *Star Trek* and the computer-assisted colorization of old black-and-white movies. Soon we will see dramatic improvements in image quality. The fact that individual details vary little from frame to frame whereas artifacts such as grain generally vary to a greater extent will allow us to identify the artifacts and remove them. And why not drop an image in, or out, at the same time? Soon it may be all but impossible to differentiate between computer-created and "real" images. When we reach that point, the line between reality and imagination will blur as never before.

We can expect technology to provide us with safeguards to validate images. Some media, such as microfilm, will be prized simply because of their immutability, and electronic audit trails and "signatures" will be developed specifically to confirm the validity of images. But given some reasonable assurance that images have not been altered, I believe that users will welcome this new environment.

Here again we have some historical precedent. Early movie goers had to learn about "camera tricks." Around the turn of the 20th century, there were reports of people fleeing from an auditorium when first confronted with a life-size image of an oncoming locomotive. There are, however, no reports of anyone being fooled twice. Today's audiences have come to relish and even demand well-executed special effects, and some forms of image manipulation have become so commonplace that they are scarcely noticed. No one is particularly surprised when, within 5 minutes, the evening news shows Dan Rather in front of the Great Wall, the Eiffel Tower, and the White House.

The integration of images into computer-based information systems has already begun, and is limited only by the available technology. Systems are now available which can handle images as simple information objects. When we have developed the capability to manage the input, storage, retrieval, transmission, and output printing of images in a distributed information network, image servers will appear. Ideally, these servers should function as "intelligent assistants", helping to locate relevant images and extract information from them. But this ideal will be difficult to achieve. For example, imagine you were handed an old class photograph and told that one of the forty students pictured was your father when he was 12. The odds are good that you could pick out your father in a few seconds, even if you had never seen a picture of him at that age. Today's best computer systems would take hundreds or thousands of times longer to "find" your father—if they could do it at all.

The problem is that images contain a lot of data—a continuous-tone color image can easily hold 100 megabytes—and it isn't easy to find the important parts and interpret them without extensive additional knowledge. We acquire this knowledge, not as a random collection of facts, but in a particular context, which we organize into an ability to understand the images we see. Although it isn't likely that image servers that "understand" images will be built in the early 21st century, we can anticipate continual improvement in their ability to sift through an "image bank" and present images that might be useful.

### Removing Technical Barriers

Images contain so much data that it is difficult to get them into, around, and out of existing information systems, and these barriers will have to be removed before we can learn how to use images effectively. The following paragraphs review the status of technology for the capture, processing, storage, transmission, and output of images, and indicate why the barriers to imaging will be gone within this decade. Existing documents can be converted to digital form by using solid-state sensors (which have largely replaced alternatives, such as flying-spot scanners, for most applications). Solid-state sensors are generally made from silicon, which converts the photons of the incoming image to electrons. The electrons then provide a signal which can be processed, stored, or transmitted to remote locations. Solid-state sensors can be made as either linear or area arrays.

*Linear arrays* are designed to scan a document one line at a time, as the document moves perpendicular to the length of the array. Systems using linear arrays are slower than those using area arrays, but linear arrays are inherently cheaper to make.

Input scanners using linear arrays can be designed with the array either spaced at some distance from the object or essentially in contact with it. In the first alternative a lens is used to form a small image of the document. Because of the small size of the image, the required linear sensor is also small, and it can be fabricated on a single piece of silicon. Linear arrays with 8000 picture elements (pixels) are available commercially.

Contact linear scanners consist of a compact illumination system (often a row of light-emitting diodes), a compact one-to-one imaging system (often a row of gradient index rods), and a full-width row of pixels formed by aligning several shorter linear arrays. It may become possible to develop lower-cost contact sensors by forming a single, continuous linear array of pixels in a layer of silicon deposited on a large substrate. At present, imperfections in such materials limit the performance of such devices, but techniques such as laser annealing may solve this problem.

We can expect linear arrays with adequate resolution for most applications to be available at reasonable cost. In most applications they will be cheaper than area arrays, and at high resolutions they may be the only practical choice.

*Area arrays* are two-dimensional arrangements of pixels, designed to capture an area of an image simultaneously, the way film does in a camera. They therefore can capture an image faster than linear arrays, which is particularly important when dealing with a dynamic scene rather than a static document. The image information generated in the center of an area array must be moved to an edge for processing, so these arrays must be fabricated on single-crystal silicon. Because of their large device area, few of them can be fabricated on each silicon wafer, and there is a higher likelihood that a given sensor will be defective, either because it intercepts a defect on the original wafer or because dust or debris falls on it during processing. This means that area arrays are considerably more expensive to produce than linear arrays, but in some cases there are offsetting factors when the costs of the full system are considered. For example, linear sensors require a more stable transport system in order to provide an acceptable image.

The technology needed to manufacture area arrays is in many ways similar to the technology required for solid-state memories, so progress in both areas has proceeded at a similar pace. A digital camera using an area array with 1.4 million elements is manufactured commercially, and 4-megapixel arrays are available.

Area arrays having appreciably more than 4 million pixels involve some difficult tradeoffs. In order for the manufacturing yield to be acceptable, the sensors must be small; but increasing numbers of pixels at constant sensor size implies smaller pixels. This reduces the number of signal electrons a pixel can hold, so signal-to-noise ratios adequate for good images become difficult to achieve.

### Image Processing

Creating, enhancing, and manipulating images, and compressing and decompressing them for efficient storage and transmission, are increasingly important functions of information systems. And because of the large amount of data which images contain, reasonable processing times require lots of processing power. In the near term, pipeline processors and parallel processors provide good solutions, and neural computers soon may be useful in some applications. In the longer term, optical computers or their equivalents should become practical; even farther out we can at least think about molecular computers.

*Pipeline processors* are like assembly lines in a factory. Data move from one processor to the next, and at each processor a specific operation is performed. The data are therefore processed much faster than if a single processor had to perform all the operations before accepting the next piece.

*Parallel processors* are like work teams in a factory, each team performing all the operations necessary to achieve the desired result. Productivity comes from using many teams, all working in parallel. This approach works particularly well with images, which can be divided into subareas, each of which is assigned to one processor which performs the desired operations. Parallel systems, therefore, hold the promise of greatly increasing image-processing capability, enabling users to manipulate and integrate images into documents just as readily as they do now with text and data. Imaging workstations containing parallel-processing boards which manipulate images at more than 1000 MIPS (millions of instructions per second) are now available at moderate cost.

*Neural nets* consist of many identical processing elements each of which is connected to others in a web-like arrangement. An important feature of a neural net is that each of the interconnections has an adjustable positive or negative "weight," which allows the influence of each interconnection to be changed. As a result of this peculiar structure, a neural net is a fundamentally different kind of information processor. Whereas conventional or parallel processors are algorithmic, requiring a set of well-specified procedures to perform a function, neural nets are "trained." For example, the front end might be presented with an input, producing some result at the back end. Then the interconnection weights would be adjusted to minimize the difference between the actual output and the desired output. After a number of such cycles with different inputs, the output of the net can be used to distinguish among the various inputs. In essence, the neural net has taught itself (like a baby) to differentiate among the objects in its environment.

At present there is no adequate theory for neural nets, so it is difficult to judge where and to what degree they might be useful. They may prove most useful in imaging applications that involve recognition or detection, including situations in which images are fragmentary or objects are viewed from different angles. Systems using neural nets for optical character recognition are now becoming available.

*Optical computers* use photons, which have important advantages over electrons as the medium for information transfer within a computer. Because they have no mass and no electric charge, photons move faster, dissipate less energy, and don't interact with one another. These properties translate into higher speed, lower power consumption, and greater bandwidth. Unfortunately, these same characteristics make photons harder to control, but some progress is being made on emitters, detectors, waveguides, modulators, and gates. It seems reasonable to assume either that optical computers will eventually become practical or that they will be rendered unnecessary because advances in materials technology (e.g., bandgap engineering and high-temperature superconductors) will have brought electronic computers to about the same level of performance with less effort. In either event, a considerably higher level of performance will have been reached.

*Molecular computers* consist of arrays of molecules. Each molecule acts as a multilevel switch and cooperates with its nearest neighbors in order to compute. A possible model for a molecular computer, proposed by John von Neumann, is the "cellular automaton." Von Neumann showed that if each cell (molecule) in a planar array is connected to its nearest neighbors, and if it can "switch" into a number of states depending on the states of its neighbors, then any problem that can be solved by computation can be solved by such an array. Moreover, it has been shown that entering the appropriate initialization program into one corner would cause information to propagate across and program the whole array, which would then proceed to solve the problem. Can we in fact design the right kinds of molecules and place them in the appropriate arrangement and environment? Maybe.

My point is simply that there are alternatives to the conventional computer architecture for the efficient processing of images, and that there are reasons to be optimistic that processing power will advance as quickly in the future as it has in the past.

### Image Storage

It has been predicted that the information output of business and government will grow by nearly 5 percent per year through the mid-1990s. By that estimate, the total number of new documents produced annually will reach 1.6 trillion in 1995—a 60 percent increase in just a decade. Even without the slightly cynical view that an existing document is more likely to be duplicated than destroyed, we are clearly facing a flood of documents that will place enormous demands on storage technology. This mass of stored information will naturally complicate the problem of finding and using the information we actually need. The incorporation of images is likely to make this problem both easier and harder to solve. Images will provide us with a new set of tools with which to browse through the system and find the data we want. Yet images will complicate the problem in two ways. First, as objects to be stored they will use up a lot of storage capacity. Second, a particular image is likely to have a large number of attributes, any of which might be used to store or retrieve that image.

Computer systems use a hierarchy to manage information storage:

*Primary storage* is fast solid-state memory contained in the processor.

*Direct-access storage devices* are connected to the processor, and provide fast access to stored information. Typically these devices use magnetic or optical disks.

*Sequential-access storage devices* store information on cassettes of magnetic tape or microfilm.

As one moves down the hierarchy, access time goes up, storage capacity increases, and cost per bit of capacity goes down.

*Magnetic and optical disks* store data on circular tracks; they provide reasonably fast access to the stored data, but they are limited in capacity by the disk diameter.

Magnetic-disk storage offers faster access, because of the lower mass of the magnetic read/write heads. This advantage will decrease in time as the mass of optical heads comes down. Another advantage of magnetic disks is a higher data-transfer rate, which is due to the higher in-track recording density in magnetic recording. This advantage is likely to persist.

The disadvantage of storage on magnetic disks is that data cannot be stored as compactly as on optical disks. This is due to the fact that the signal read from a magnetic disk is derived from the magnetic fields recorded on the media, and its strength depends on the amount of media passing under the reading head. In practice, this means that tracks on magnetic disks must be wider than on optical disks and so the center-to-center track spacing must be greater. If the signal level read from magnetic disks continues to scale linearly with track width, as experimental evidence indicates that it does, then magnetic-disk storage should continue to reduce but not eliminate this disadvantage.

Multiple-disk storage devices are available for both magnetic and optical direct access. Magnetic systems use disks stacked on a common spindle within a sealed container, with heads addressing each side of every disk. Optical systems use jukebox arrangements to bring disks to a small number of optical heads. Once again the optical system has slower access, but such systems can store over a terabyte (a million megabytes) of data.

Several R&D groups are working on disk arrays in which many small, inexpensive hard disks are combined into a single storage subsystem. Disk arrays promise high capacity, low cost, and fast access due to the small diameters of the individual disks. The challenge lies in the data management, particularly in the event of the failure of one or more of the disk drives.

The roles played by the various magnetic and optical storage devices in on-line storage of images will clearly depend on the tradeoff between access time, data-transfer rate, image size, and cost relative to application.

*Sequential-access storage devices* provide slower access to much larger quantities of stored data. The data are stored on rolls of microfilm or magnetic tape, which are brought to a reading device when required. Microfilm systems use solid-state sensors to convert microfilm images to electronic form; magnetic-tape systems can use either multi-track linear heads or helical scanning heads for reading and writing on magnetic tape.

Microfilm systems are archival, and the stored information can be used as legal evidence. There are really two considerations here. The first is the permanence of the recorded data, both because of the stability of the microfilm image and because it cannot be altered (either accidentally or by a computer "virus"). The second consideration is that microfilm images can be read by the human eye, so the image can be retrieved even if system standards change.

Microfilm also offers fast image capture (some microfilmers run at more than 10,000 documents per hour), low cost per image, and fast, inexpensive duplication.

Magnetic-tape systems are best suited to applications characterized by frequent updating and short useful data life. Helical scan systems have high storage density and moderate data-transfer rates. Linear recording systems have lower storage density, but can have considerably higher data-transfer rates if linear arrays of heads are used.

In both microfilm and magnetic-tape systems, access time can be substantially reduced by automating the processes of locating the appropriate cassette and reading the data.

It seems likely that imaging systems will continue to use a combination of storage media to meet their performance requirements. For example, imaging information systems with archival requirements are likely to capture documents by simultaneously scanning them (with storage on direct-access storage devices so that the information can be accessed immediately) and microfilming them for archival storage. A key design feature of these systems should be that the system manages the image storage automatically, without requiring the user to specify anything other than the requirements.

### Image Output
Users prefer high-resolution soft display and "what you see is what you get" hard copy. As the use of images grows, these preferences will become requirements.

For soft display, cathode-ray tubes (CRTs) will continue to be the best near-term choice in most applications. They should be able to meet the requirements of desktop display well into the future.

For portable devices, where compactness and low power demand are important, flat panel displays based on liquid crystals addressed by thin-film transistor arrays should be adequate for many applications.

For large displays, projection systems using either computer-driven light valves or projection TV (based on extended-definition or high-definition television) will provide images of good quality.

*Hard-copy output* will be an essential part of information systems for the foreseeable future. Technology has enabled us to raise our standard of living over the last decade while using less per capita of most raw materials. A notable exception is wood pulp for paper, which we consume at ever greater rates; it seems that the paperless office is farther away than ever. As information systems gain the ability to use images effectively, pulling them from databases around the world and incorporating them in compound documents, the need for high-quality hard copies of images will grow. This need will be met by electrophotographic, ink-jet, and thermal-transfer printers.

*Electrophotographic printers* use uniformly charged photoconductors which, when exposed to an image, discharge in the illuminated areas. Charged colorant particles are then attracted proportionally to the photoconductor, transferred to paper, and fixed with heat. Some electrophotographic copiers can produce copies of textual documents that are virtually indistinguishable from the originals. Advances in this technology will provide continuous tone and color images which may eventually approach today's photographs in clarity, contrast, and color density.

*Ink-jet printers* use either shock waves generated by piezoelectric crystals or vapor bubbles generated by small heaters to propel ink droplets onto paper. These printers can produce documents with good resolution, although the image density tends to be low unless special paper is used. The ink-jet process can be very simple mechanically and requires little power; thus, it is useful for inexpensive and portable printers.

*Thermal printers* use a linear array of small heaters in a printing head to drive dyes, inks, or waxes from a donor sheet onto a receiver sheet. This process is relatively slow, because the heater elements require time to heat up and cool down, but it is capable of high resolution because heads having many closely spaced heating elements can be produced by methods used to make thin-film electronics. Even higher resolution can be achieved by using lasers to transfer the colorants, but this process is even slower because the ability to write an entire line at a time is lost.

An important issue for hard-copy output is the desirability of color images in the business environment. Historically, the role of color has generally been underestimated before its introduction: "Who needs color in magazines?" "Will anybody pay for color TV?" "Who wants color in newspapers?" The ability to generate electronic documents in color, either by scanning color originals or by composing them on PCs or workstations with color CRTs, is now spreading rapidly. This will drive the demand for color hard copy.

### Image Transmission
Of all the elements needed to provide an effective electronic-image information environment, transmission is the one most in doubt. The problem is not so much a matter of technological development, although clearly some is required, as a matter of priority and the will to proceed. Two things are required: a commitment to a standardized, high-bandwidth, seamless communications network, and the investment necessary to make that network widely available.

The present patchwork of proprietary and special-purpose networks simply does not provide easy access to information. It is too cumbersome to be useful. The present facilities make no more sense than allowing each public utility to choose the voltage and frequency of the power it supplies, or allowing each railroad to use a different gauge. It will take a considerable investment to create the network that will be needed, and industry alone is not likely to have the money or the incentives to provide it. But electric utilities, telephones, and the interstate highway system all have required either direct or indirect federal subsidies, and each has proved to be a good investment. A national information network requires similar priority and will be as worthwhile.

## The Information Environment of the 21st Century

Given that the technological advances discussed above will in fact be achieved, what will be the result?

In the late 1880s, Edward Bellamy made some predictions about the year 2000 in a novel titled *Looking Backward*. Among other things, he predicted that consumers would be able to listen to music in their own homes. They would, he said, simply dial up the local symphony hall on one of those newfangled telephones, and listen to selections played live over the receiver. Bellamy's work was published only a few years before the phonograph and wireless transmission were invented. That's about the same amount of time we have left before the year 2000.

### The War Room
What follows is an example illustrating an integrated information environment, transparent access to data and systems functions, and the impact of such systems on competitiveness.

Joanne Williams had just finished reviewing the first-quarter sales of ChowHound canned dog food when Andy called.

"Joanne, we have a problem in Eastern Ohio. Can you meet me in the war room?" The concern in Andy's eyes was apparent even on the screen of Joanne's phone.

"Sure, Andy, I'm on my way." Joanne blanked the screen of her workstation and headed for the door.

The war room was located on the second floor, and its doors opened onto a balcony which overlooked a large, semi-transparent screen. Joanne had arrived before Andy, and the screen was dark.

Andy pounded through the swinging door, staff assistant in tow. He didn't waste any words in greeting, but went right to the railing, looking at the screen.

"Northeastern Ohio map," he said. The screen glowed softly, then a map appeared.

"Change in market share, last 24 hours." The map went mostly yellow, except for a large dark red patch centered on Canton and spreading into Akron and Youngstown.

"Look at that," he said. "We've lost the whole market to Attaboy, basically in the last four hours. What's going on?"

"Same plot, entire US," said Joanne. The image on the screen shifted to a map of the whole country. Only Northeast Ohio contrasted in vivid red.

"That's really strange. Maybe it's a market probe of some sort. Have you checked for specials, advertising, inventory problems?"

"It looks too drastic for that, Joanne, but let's have a look." Andy barked a few more commands.

Graphs of prices in the last four hours, advertising hours in the last week, and inventory for both Attaboy and ChowHound went up on the screen, but there were no surprises.

"Let's see their most recent TV ad," said Joanne. Thirty seconds later they were convinced that wasn't a factor either. The room was quiet as they thought the problem through. They could see the price graphs slowly extend to the right as customers in the Canton area passed through the laser scanners at checkout counters.

"How about coupons?" asked Joanne. "Plot coupon redemptions versus time."

There it was—a large upturn in customers using Attaboy coupons in the last six hours.

"Percent of customers using coupons versus time," said Andy. This graph also showed the recent upturn.

"Something's happening all right, but that can't be the whole answer—less than half of the buyers are using coupons, and Attaboy's got the whole market."

"Yes, but we're onto something," Joanne observed. "Let's see two graphs: percent Attaboy customers using Attaboy coupons and percent our customers using Attaboy coupons, both in the last six hours."

Data on customers, identified from their charge and debit

cards, sorted by their prior brand preferences, appeared on two graphs.

"That's it, Joanne," said Andy, "Nice guess!"

The graphs showed no coupon use by Attaboy customers, heavy use by ChowHound buyers.

Joanne understood what was happening. "Looks like a selective mailing of coupons to our customers. They're wiping us out without hurting their profits."

"Right. What do we do?"

"Well, if we do nothing, they may try the same thing in a larger market. So we have to respond, either offensively or defensively. Offensively, we could do the same thing to them, either in Canton or in a larger market—say Pittsburgh. Defensively, we could start a TV and a newspaper ad in Canton this evening, stressing the quality of our product and implying that our competitors have to slash their prices to sell anything. And we could run a sweepstakes. Anyone buying in the next two days is eligible for a week in Hawaii. We could guarantee a winner in the Canton area."

"Projections," ordered Andy.

The screens showed graphs of predicted corporate earnings under the two scenarios, based on corporate models of customer behavior. They slightly favored the defensive strategy.

"The defensive strategy is the way to go, Andy. It keeps Attaboy guessing. The offensive option tells them too much about our ability to analyze their actions."

"Okay," said Andy, "how fast can we get moving?"

"Well, we can get the ads on TV this evening and have the sweepstakes flyers in the morning papers. We've probably missed the mail window, so the flyers won't be in the mailboxes until the day after tomorrow."

"Guess that'll have to be good enough, then." On his way out, he said: "And nice job, Joanne."

### Quasicorp
The next example illustrates the impact of integrated information systems on business relationships.

The phone rang just as George Bemis took a gulp of coffee. His monitor flashed "Ed Campbell." In the upper right corner of the screen a file cabinet appeared, indicating that the data he would need to field the call were being retrieved.

George took a moment to swallow, then punched "Receive."

"Bemis here. How's it going, Ed?"

"Not bad. Say George, yesterday we sent some stuff over to you. We've found a terrific product opportunity, and we're forming a corporation to go after it. Usual arrangements: we'll disband when the ride is over. I'd like you to sign up for manufacturing, but I need an answer pretty quick. I want to call a sign-up meeting for eleven."

George scanned the figures on his screen again, but he had long since absorbed the crucial information.

"Yes, I know, Ed, but I still have to look at it for a while. We've got to do a lot of shuffling to meet your schedule, and frankly the financials don't look so terrific. You're squeezing us pretty hard."

Ed brought up the financial terms of the contract on both terminals.

"I can't help you out on the price, George. That's firm. I might be able to help you a little on the residuals, though."

"Ten percent?"

"No, try five, George. I'm pretty thin, too." The monitor showed Ed's changes in red as he entered them on the contract.

"Okay, Ed. I'll crank it through. I'll get back to you by nine," George said.

"Thanks. I hope you'll join us." Ed's name disappeared from the monitor.

George wanted to take the deal. Ed Campbell had consistently put together good partnerships. And the financials really weren't so bad, especially with the better residuals. But if the product was a complete bust George could lose some big money, and if it was an instant success the volumes would go up fast and George (who specialized in fast, low-volume production) would have to hand off the manufacturing to a high-volume specialist. The increased residual profits would help after he dropped out, but clearly his profit would be better if he could hang in for a while.

George decided it was time to call Alex, an external product consultant based in Toronto. George couldn't be an expert in all the markets his products served, and Alex was one of the best in this area. Of course, talking to Alex had its risks too—everybody in the market talked to Alex, and he made his living by selling what he knew. But yesterday George had decided that the benefits of talking with him outweighed the risks and had sent him a scrubbed version of Ed's proposal. George shut down his internal links—no need to be careless—and called Alex, whose name blinked only briefly before he came on.

"Morning, George," said the Canadian.

"Hello, Alex. Let's have a once-over, usual terms."

"Fine, George. Well, here's what is in the market now."

Alex went through half a dozen products, citing prices, performance, and sales volumes in each of the world's markets. As he described each one, its image rotated on the terminal above the graphs and charts. Alex finished with a brief summary of what seemed to sell. Then Ed Campbell's product came up on the screen, slgihtly fuzzed so that the fine styling details were lost.

"Your proposal does provide some new functions, but I don't think they'll help you much. But the compactness should make it a winner, and the design makes it look even smaller than it is. Lycon did a study of the market about two months ago, and they came to the same conclusion: small size will sell."

A bound copy of the Lycon report dropped out of George's printer; his terminal showed page 31, a bar chart demonstrating that reduced size was the most desired new attribute, particularly in United Europe.

"Bottom line: if you price it right, you should do okay. I can estimate volumes if you give me a rough dealer net."

"Not my game, Alex. Do you expect any new entries?"

"Well, there's one I've heard about that could be a factor. The rest is junk."

A new model appeared and began to rotate. It had also been blurred, and George couldn't read the nameplate.

"It's not committed yet, and it may not have the appeal of yours. But anyway, George, if it does hit the shops it could mean a shared market, and the lower volumes would mean a longer run for you, right?"

"Well, thanks, Alex. Let me know if anything comes up."

George shut down the external links and composed an acceptance note, to be sent at 8:50. Then he went out to warm up his coffee.

At 11 the phone rang again. Ed called the meeting to order and seven windows opened up on George's monitor, showing each of the participants.

George noted that he was working with four of them on other products. Two of them were not fluent in English, so Ed kept the introductions short. Then an icon indicated that Legal-Ease was on line from Philadelphia to monitor and record the agreement.

"We've got a deal!" said Ed. "Here's the final copy of the contract."

George's system received the document, then presented it page by page, with Contract Advisor noting all changes in red. They were minor. Coming to the principals' page, he noted that Federated Insurance had signed up for three elements: financing, receivables, and warranty coverage. Nearly everyone had negotiated some changes in the financials. George tucked that away for future reference.

A signature page came out of his printer. George signed it and slipped it into his Out slot. Legal-Ease went off line and Ed said: "Okay, we're in business. I'll be back to you all on Thursday for a progress check."

The screen went blank. George turned and found that his coffee was still warm.

### Panther

This example illustrates on-line computer modeling and simulation, user testing of product simulators, tele-travel, and artificial environments.

Peter walked into the conference room behind Linda. He stopped to straighten his tie in the wall mirror before heading for the table.

"Traffic was bad this morning."

"It sure was," she replied. "Maybe someday we'll be able to hold these meetings at home rather than fighting through that stuff. You all set, Pete?"

"Sure, go ahead," he replied, opening his attaché case.

Linda hit the Ready button, and the warning light began to blink. They must have been the last to arrive.

A low beep sounded and images appeared at chairs around the table. There were eight people, counting Linda and Pete, from three different cities.

"Ummm, good morning," said Braxton. It was clear that something was wrong. "Three things on the agenda today." An agenda dropped into the copy tray in front of each chair. "First, we'll review the rest of the results on the prototype, then we'll look at the design implications of any necessary changes. Finally, we'll walk through the factory to see how they're coming."

Pete glanced around the room. The second agenda item meant that the tests had uncovered problems.

"Doris, why don't you review the prototype results?"

Doris Mathews leaned forward and began. "Last time we agreed to try the prototype simulator at seven sites. Subjects at each site were first shown a model of the car for reaction." A small model of the Panther appeared in the center of the table, spotlighted and rotating slowly.

Problems or no, thought Pete, that is one terrific-looking machine.

"Response was overwhelmingly positive." Whew. Pete suddenly realized just how nervous he had been. So much rested on the results of these tests.

"The subjects were then asked to test-drive the simulator. None objected or found the experience artificial. In fact, several subjects developed sweaty palms during the test drive."

Nervous chuckles around the room. Apparently, he wasn't the only one who had been concerned.

"Comments on acceleration, handling, and ride were generally favorable, with the usual outliers. The only statistically significant complaint was that the visibility is poor."

"What?" said Pete. That was the last thing he expected to hear about the Panther.

"The visibility is poor, and that seems to be associated with the bulges of the fenders above the wheels."

"Come on, Doris. The fenders aren't high. The hood is low! This car has terrific visibility!"

"Pete," said Braxton, "the bulges have to go."

"The bulges can't go. We need room for the wheels, you know. The hood would have to come up."

Linda groaned. "Do you know what my engineers went through to accommodate that low hood?"

"Let's see it," said Braxton.

Pete stopped the rotating model, enlarged it, converted to wire frame, raised the hood, put a forward-slanting scoop on the front end, and put back the outer skin.

"Let me check the wind tunnel," said Linda. Luminous blue lines appeared and flowed around the black model, showing a few areas of turbulence. These disappeared as the supercomputer gradually changed the car's shape. Linda then started the model rotating.

"The drag coefficient hasn't changed much," she said.

"I like it," said Braxton, to murmurs of assent. "Does it give you any problems, Linda?"

"No, actually it should be easier with the extra room. We may even be able to cut the costs a bit."

"Okay, Pete?" said Braxton.

"Well, I guess we have to go with it. But I like the original design."

"Then we're decided." Braxton smiled. "Doris, have the new design checked out for next morning, and Linda, please confirm the engineering. That covers the first two agenda items. Now, since most of you have never been to the Rolling Ridge site, I've asked Hari to give us a quick tour. Hari?"

The table went dark, dropped away, and a wrap-around image enveloped Pete and Linda. They were several thousand feet in the air, and Pete had a firm grip on the arms of his chair.

"We're facing north, and that's the city of Rolling Ridge just ahead of us. You can see the interstate just north of town, and the railroad is just beyond that. If anyone cares to stay after the meeting, I'll be glad to show you around the town. The plant is to the west, and we'll head over there now."

The helicopter flew over to the site, circled it twice, and landed on a bull's-eye on the near side of the parking lot.

Asa Deverian, the plant manager, was waiting near the main building. After a brief welcome, he escorted the group inside. It was a single high bay area, not as large as you might expect, occupied by rows of crates and construction materials. Overhead cranes were shifting things around. Lou asked to see the forming areas, so they moved to a main aisle and headed west.

Not much going on. Pete had expected to see more activity.

"While we're walking over there," Asa said, "let me show you how the area will look in about three months."

The scene blurred. Then they were moving through a simulated factory, with the equipment clean and neatly arrayed and the people working efficiently.

"All of us in Site Management have been through the plant many times, studying every aspect of the workflow. We've been able to make hundreds of changes that will minimize problems later."

Or maybe, thought Pete, that's a problem too.

"Well, here we are," said Deverian. Before them stood several massive forming machines flawlessly making body parts. Hoods, in fact. Sadly, Pete noted the low, graceful lines, which would change when the simulator caught up.

The scene blurred, then came back to reality. Only one forming machine stood in place, and boxes and parts littered the area. Several workers moved among them, and one of them nodded to Asa.

"How does the schedule look?" asked Lou.

"Well, I think we'll make it. At least, I don't expect any problems."

"Fine," said Braxton. "Hari and Asa, thanks for the tour. Our next meeting is in two weeks, and we'll be back to see your progress. Well, I have got to get back. Anybody who wants to stay can look around, and maybe have a tour of downtown. See you all in two weeks."

Pete and Linda decided to cut out. They thanked Asa, the light in the conference room went on, and they gathered up their stuff.

"Tough day," said Pete.

"Yes, but a nice recovery, Pete. Unfortunately, you've given me a load of engineering changes to make. I'd better get at them."

"Well, anyway, I hope the traffic is better this evening."

"It's got to be better than coming back from Rolling Ridge."

## Conclusion

"The War Room," "Quasicorp," and "Panther" described imaging environments with transparent access to distributed databases, extensive modeling capabilities, multilingual and international communication, and realistic scene simulation. These examples pose several provocative questions.

*Will imaging technology be used to provide alternative realities which will be attractive alternatives to the real world?* It would seem that this is inevitable. Television and motion pictures already serve that purpose, but at great costs of production, and the number of hours the average person spends in front of a TV set clearly shows that the demand is there. New imaging technologies will simply make alternate realities more "realistic," more adaptable to individual interests, and eventually more interactive. At some point, image synthesizers will assume roles analogous to those of music synthesizers, with a few artists expanding horizons, a number of users who will create their own environments from existing components, and a majority who will enjoy the results.

Businesses will use imaging technology to humanize the workplace, making it more engaging and therefore more productive by drawing upon the experiences and the interests of the workers. Each enterprise will be able to tailor its information environment to help its employees understand their responsibilities and optimize their performance. Perhaps each person will be able to define the way his or her job is experienced. For example, if you were able to watch over the shoulders of workers in an insurance company as they did their jobs, you might find one person studying the flow of freight cars through a rail yard, another in the midst of a Civil War battlefield, and a third designing the logistics to support a colony on Titan. After lunch, each worker might choose a different environment.

*Will we be able to use this new imaging environment to model the real world, rather than an artificial one, in order to predict the consequences of different actions?* Although imaging will continue to provide the ability to quickly and easily see the results of very complex calculations, there are fundamental limitations to our predictive capabilities. Recent results in chaos theory have demonstrated these limitations, even in cases where the governing relationships are well defined and no randomness is involved. The problem is that apparently simple nonlinearities cause such strong dependence on initial conditions that systems which appear to be the same (that is, whose differences are too small for us to measure) soon diverge and become dissimilar. As surprising as it might seem, although we know Pluto's present location and the laws governing its motion, it now appears that we cannot predict its position millions of years from now. Technology won't enable us to predict the future. But imaging information systems will provide us with an unprecedented ability to understand our world as it is and as it might be, and to build a future which each of us will find more interesting, more productive, and more fun.

# *Information and the Human Race*

# 10

# A World to Understand: Technology and the Awakening of Human Possibility

*Derek Leebaert and Timothy Dickinson*

Ophelia: We know what we are, but we know not what we may become.
*Hamlet*

Why was Athens, between 510 and 420 B.C., able to produce the highest concentration of human achievement in history? The backward, inefficient little town on the Aegean unleashed unequaled literary forms, rigorous philosophical inquiries, and profound studies of history. One achievement begat the next. Yet Athens had a far lower proportion of citizens educated at what we would consider the high school level than does present-day Athens, Georgia.

There was nothing biologically different about the Athenians, many of whom were not even Athenian by ancestry. And around the same time, in the fifth century B.C., a similar surge of creativity appeared in China. In both instances, there was a freeing of the consciousness. The sense that with the acceleration of present achievement almost anything was possible swept through society in what economists call "the demonstration effect." The knowledge that certain extraordinary things had been done by people of genius stimulated those of less genius— or less-realized genius.

Why haven't other superficially marginal, disorderly cities blazed like Athens? Will technology soon emancipate and empower us so that we will take such outpourings of originality for granted? And why does one person's scientific achievement suddenly spur several others—not even knowing about the original breakthrough—to suddenly solve the next step of the problem in different ways?

In Athens—a city having the male population of Dayton—
Pericles, Sophocles, Euripides, and Aristophanes rubbed shoulders in the street. In more recent times, particular high school
and college classes have been bursting with talent. For example,
Bowdoin College's class of 1825 produced both a US president,
Franklin Pierce, and the greatest American novelist of his generation, Nathaniel Hawthorne, as well as other luminaries.
With all due respect for Bowdoin, graduates of such distinction
do not show up every year. Coincidence is no answer. There are
too many correlations for the result to be random.

Up to the Spanish conquest, the great Central American cultures had wheeled toys but never developed wheeled vehicles.
And before the bridle, people all around the world rode and
drew plows by tying ropes around their horses' necks. The
harder the horse worked, the more it was throttled—classic kinetic feedback. Why wasn't it seen that a small adjustment
would prevent such strangulation? The stirrup appeared only
in the ninth century, transforming the use of the horse in multiplying mankind's capabilities.

Solutions that are perfectly obvious in retrospect are incon-
ceivably obscure up to the point of their discovery. The problem is to shake off preconceptions. How do people learn? And
how do they unlearn? Few societies have had an exponential
access to information and an exponential capacity to enlarge
upon it, and Western industrial society is the only one to have
had so long a ride.

It is unlikely that Athens resulted from cosmic purpose. It
was just another place, as it is now. But something happened in
that quarrelsome, warlike little port city that distilled a century's
worth of incredibly talented, reasoned, and enlightened imagination out of essentially random genetic and cultural material.
Certain constraints were suddenly removed. Athens' achievements are all the more impressive since we know of only a fraction. For example, long lists of lost plays have been discovered.
It is known that Sophocles wrote 123 plays, of which seven survive. As the classical historian A. R. Burn observed sadly, "We
have from the ancient world, at best, about as much as you'd
get from a badly bombed record office."

Yet if one believes in the immense possibility and variety
of human interaction with the environment, as Marshall
McLuhan argued, all that is really needed to ignite an illumination like that of Athens is a spark—a spark which, when it

falls on the right medium, kindles a train of energy. Just so, probably, was life started by some well-placed proton cruising through the rich molecules of the primeval sea. But there is a great spectrum of hospitality across environments. And that is probably the most frustrating observation that one can make in life.

Moreover, there is a continuous spectrum of the transmissability of talent. Virtually every baby grows up to speak the language of its native land; virtually no painter becomes as great as Rembrandt, and none has ever become like Rembrandt. In fact, transmitting such talent in painting is infinitely more difficult than transmitting it in literature. Great schools of Mogul painting, for example, originated in tiny areas.

Among mathematicians, it is not amazing to discover the occasional superman (such as Newton, who conceived calculus in his twenties). No one is surprised to discover musical ability in adolescents. In fact, we expect great composers to have produced something startling before the end of their teens. Certain capacities are likely to manifest themselves in youth; other capacities take time to unfold. Neither Aeschylus nor Shakespeare produced great plays at 20, although it is likely that hints of their gifts were emerging.

Therefore, in what spirit do we try to awaken capacities in our children? Some expectations will not be justified until they are in their thirties; others clearly justify themselves, or fail to do so, in the first few years. The objective is not to score well on tests per fortnight. Instead, the challenge is to initiate achievements that will sustain themselves over a lifetime.

### A New Intensity

The theme that is most intense through the rise of Athens and the awakening of each child is that of *de-passification*—both of knowledge itself and of ourselves before it. The computer increasingly provides feedback and organizes interrelated information while continuously expanding the frontiers of interrelation. It expands the range and the scope of designated participation. It awakens possibilities of creation among the many, rather than just the minorities of past ages. After all, most of human achievement has been the anonymous assembly of discoveries and insights, from proverbs to building techniques, and their equally anonymous loss. The individuals within the great faceless creating mass of the human race can now relate to one another directly in real time, rather than over trade routes and centuries.

The computer de-routinizes, and routine is a sedative to the mind. The general relegation of trivial tasks to the mechanical periphery is perhaps the most explosive phenomenon of computerization. Millions of environments find their interior obstacles leveled or, perhaps more accurate, cut away like a hillside to let hurtling achievement beget achievement.

First, the computer increases the gradient of intensification, accelerating history as history was accelerated in Athens' surge from obscurity in 520 B.C. to primacy by 420 B.C. Second, the computer can intensify the rigorous, formalized disciplines that are at the heart of intellectual life. Plato emphasized such tests of seriousness, carving "Let no one enter who is unskilled in geometry" over the door of the Academy. The analytical engine can rapidly subject abstract speculation to an infinite number of empiric tests. Third, and most important, the computer can enhance the sense of possibility, giving us a sense that there are no fundamentally unconquerable categories. People now expect to be fascinated by information and to thread their way unastounded to its most applicable areas. When a new phylum of living beings was discovered a decade ago at the heat vents of the deep oceans, the reaction was hardly intense. The information revolution had already generated such a sense of the vastness and prodigal fertility of the biosphere that this genetic adaptation, in a real sense stranger than life on Mars, was just assimilated to what had gone before.

There were only five surgical operations for the eye from 300 to 1600 A.D., so there was hardly a need to keep abreast of them through the professional literature. People are now overwhelmed by the needs of specialization; however, they can sort for information through ever-wider data banks, and they can take numerous shortcuts. One door flies open and the next door opens before we can put a hand on the knob. Along with the availability of knowledge has come the availability of the meta-skills that maximize the use and the exchange of that knowledge. The ability to concentrate and disburse information is becoming ever greater.

The burdensome aspects of information and knowledge are diminishing at the same time. Memory was once a master skill. Men of genius, such as Bruno, poured their energies into the training of memory as an occult art. Pre-literate societies were served by bards with hundreds of thousands of lines committed to memory. More and more, however, people found themselves seeking insight and process. Data need not be kept in mind; they can be looked up. A generation ago, James Blish rightly observed that to immortals memory might prove "a Greek gift," burying mind in data. The dynamization of information by turning it into a machine process rather than an accumulation of inert files—the book, the magazine, the document—has given wings to inquiry.

Many Athenian citizens participated in debates. Aristophanes could ask "What then, Sophocles?" Socrates could walk a mile or two for a word with Thucydides. In 17th-century Europe, however, a professional intellectual correspondent such as Nicholas de Peiresc could make a distinguished career of just maintaining connections among his peers. A month would pass as letters were exchanged between Paris, Oxford, and Bologna. The intellectual environment is once again more like the vital Athens of the agora and the Academy than the Toulouse of Peiresc.

The computer permits discourse to be more concentrated yet more planetary, more varied yet with more depth of knowledge, more convenient yet more swiftly recorded (if not yet more original) even than it was in Athens. Yet technological facility does not ensure intellectual hunger. There is the danger that things will be looked at with insufficient surprise. It is possible to take the edge off the sense of wonder, as Constant Lambert explained in his essay "The Dreadful Popularity of Music." Music on elevators and in shops becomes "as wallpaper," as McLuhan said. That is a result of educational initiatives in the larger society. When Harry Augustus Garfield said that the best education was a log with Mark Hopkins at one end and a student at the other, the crucial medium was not the log; it was the quality of concentrated discussion.

Except for the log and the highly passive blackboard, virtually all media are now shifting toward a dynamic technology. (In fact, the blackboard itself should be an instrument for a continuous loop of discussion in a roomful of people asking questions.) One might want the latest newspaper displayed on one's screen, with ten-minute updates. But the computer also poses the danger of tempting the student not to ask questions.

The possibilities of everyone's having the opportunity of a dialogue with Socrates, rather than just a Socratic dialogue, are technical as well as economic. Originally the best that one could do was to ask Socrates a question. Now one can imagine a Socratic program that will engage the user with a great range of Socrates' counter-questions and methods of inquiry—what Socrates would likely have said, or what he did say in response to similar questions. (A sufficiently hospitable system would need gates; otherwise 90 percent of the questions could be answered from the encyclopedia, or would be unanswerable.)

Eric Hoffer noted that "most talent is wasted." Virtually all imaginative capacities are unawakened. And this is saddest among children, each of whom has a vast capacity for learning. Einstein claimed that all that was different about him was that he kept asking the questions that everyone asked at age 5. Bach and Handel were born in the same year, but we can assume that countless other young talents in 18th-century Germany were unmobilized. And what about the lost brilliances of other years? If three-year-olds learn German in Germany, can we ensure that ten-year-olds learn math in Los Angeles?

Ideally, learning means acquiring early a "second nature." An eminent anthropologist used to list a great text on carpentry in his bibliographies, because it had taught him that most clearly statable problems responded to serious thought. But learning can be obstructed by what might be called "sophisticated incapacity"—the problem of being overwhelmed by the prospect of again accomplishing something you don't understand how you accomplished in the first place. Trying to learn a second language as an adult is Exhibit A.

How can we all move closer to arousing creative achievements like those of Athens among anything like a comparable proportion of our technically advantaged citizens? Can we re-create an Academy for our children? What media will encourage argument and heighten the intensity of engagement? The Greeks were fascinated by the fact that Oriental cities had no marketplace—no agora in which to debate. In some respects we have transcended the agora with our communications media. The 24-hour electronic market in ideas is a tremendous energy provider. In an interconnected society, as in the agora, a proposition that is no more than a bumper-sticker slogan can be exploded—which literally means "jeered offstage." Better but still insufficient propositions can be refined; good can be put to work at once. What until recently existed only in the little "invisible colleges" of cutting-edge professions, such as theoretical physics, now promises to weave all humanity together.

The essence of the market economy is the fact that dissatisfaction increases along with the range of choice. The computer should be a major tool of dissatisfaction, as were the relentless inquiries of Socrates. The questions the computer suggests include these: "Am I doing this task well enough when X and Y have other approaches? Can I do it better, knowing that if I don't someone else soon will?" The knowledge of so many other competing activities is a perpetual challenge to complacency. The degree of dissatisfaction that technology will induce by century's end, especially dissatisfaction with education, is likely to be high. How will people live with an increasing intensity of inconclusiveness (e.g., propositions coming constantly under test from ever more different angles), and with the perpetual appeal to experience, which will derive from technology?

### *Teaching and Technology*

Experience with the computer can be thought of as a curve stretching from the vacuum tube to the transistor, with another curve generated around 1958 by the arrival of integrated circuits. There is not one sweep of progress; it is steepeningly exponential. Focusing on silicon, the physicist Gordon Moore stated, in what we now know as Moore's Law, that all information capacities will double roughly every year—a proposition only a few percent off, but tapering around 1995.

A crude calculation indicates that the world's written sources have increased by a factor of 20,000 at most since the great library of Alexandria. By Moore's Law, we would surpass that achievement in less than 15 years.

People are now using a still very little altered pattern of mathematical learning and conceptualization. We should be able to educate children to a vastly higher level of mathematical confidence. When a doctor plays with his calculator to show the incidence of a disease, the curve that results is identical with kinetic and ballistic ones. The Fibonacci sequence is to be found threading its way through the formation of thunderheads and the growth of snail shells. Mathematics is not abstract; it is real—the grammar of the universe. Just as children master the game of language and delight in its sheer possibility for play, the implicit power to make worlds that lies in the ability to manipulate these symbols can surely reach their heart.

In the late 1960s, Seymour Papert took two languages developed earlier in the decade—Lisp and Joss—and combined some ideas from the former with the user-friendly aspects of the latter. The result was Logo. In a study supervised by Papert at a school in Lexington, Massachusetts, children of 10, 11, and 12 were soon using Logo to write serious computer procedures—such as programs to turn English sentences into Pig Latin.

Imaginative use of the computer can be thought of as mapping the immensely complex $N$-dimensional manifold of the computer's capacities and objectives onto the much vaster and harder to realize manifold of human capacity. Computer resources are on continuous call. They are less than human, but they do not stall, they do not tire, and for good and ill they do not soar and scatter like birds. The computing process is the next step in the long path from mnemonic knots through writing and mathematical notation—in working the transient brilliances of mind with the dependable fabric of the external world.

The hunter's bow probably evolved from musical instruments. How much of mathematics and logic evolved from simple argument? The human ability to explore and envision is logically antecedent to marshaling the complex skills of full practical achievement. Not for nothing do so many scientific pioneers read science fiction: "In dreams begin responsibilities."

At a certain point in the late 1960s, Alan Kay had just finished doing some work on an early desktop computer called the Flex Machine. He loved it, but its user interface repelled everyone else because there were too many commands to remember. He contrasted this experience with the Flex Machine to Papert's work with children. If children could master such computer work, then we needed to think about computers in a new way. Computers had been perceived in terms of trains: you caught the train, the train did not catch you. The computer is a continuous instrument, unlike the discontinuous service of a railway. At best, he had been thinking about the liberating experience of the computer in terms of advancing us from the train to the automobile. Stanford professor Doug Engelbart's metaphor was of a vehicle that would take you through space—

one that could even be flown like an airplane (or, considering its great abilities to do stupifyingly repetitive work, driven like a submarine). This was a liberating vision.

One of the unfortunate hidden assumptions fostered by the metaphor of transportation was that a young person needed to take Driver Education to use a computer. Papert demonstrated that the computer was not at all an adult vehicle. It was a device that could reach out to children if only the environment could be improved. "Driver Ed" implies responsibility to others. However, with all great primary skills, the most demanding responsibility is to one's own possibilities—what Jesus meant in the parable of the talents.

Papert's achievement made Kay think further about the tools (vaguely assumed to be generically adult) that were used in early childhood, such as pencil and paper. The use of these tools by children was deemed to be "playing." Even small children had access to books, paper, and pencils for their scribbling. What would happen if they had access to computers?

Papert's work was the first of three exciting developments Alan Kay noted in 1968. The second was the first flat-screen display—a one-inch-square plasma panel developed at the University of Illinois. The developers expected to have a 5½-inch-square panel within two years. But when would the components of the sort of desktop machine that he had explored be small enough to go on such a panel's back? The result of that combination would presumably with something like a notepad, a sort of "personal" computer. With the projections of the integrated-circuit experts extrapolated via Moore's Law, it appeared that such a creation would be possible by the late 1970s or the early 1980s. What Gerald Piel calls "the acceleration of history" builds extrapolation into our assumptions. Athens, however, could not extrapolate an exponential increase in its capacity, and, in fact, had pessimistic notions about transmitting its achievements. In the end, it transmitted its attitude (as a challenge) more than its achievement (as a process).

The third development Kay encountered in 1968 was a system called Grail, an interface that was a much better pointing device than the mouse. Gesture recognition was emphasized, and the effect was similar to that of writing with a pen on paper and having the computer interpret what you wrote. Kay used it for half an hour in 1968.

These three developments suggested to Kay that the metaphor of a vehicle, which people had been using in discussing computers, was misplaced. The right metaphor would be a more prosaic one, such as pencil and paper. But also there was the dim anticipation of something more astounding: a nonpassive book that would argue with you and help you read it. The computer is not a car taking you into the world. It is a living world itself, beckoning you into it as an aquarium does. There was something in the iconic interface that offered a whole different way of dealing with the machine than just typing in streams of symbols.

Together these three ideas foreshadowed the computer's power to elicit enthusiasm from humans in general, and from children in particular.

There was the question of extending the power of symbolization—saying things with tremendous economy—from numbers, characters, and letters (which gave us written memory and arithmetic) and the assignment of representative functions (which yielded algebra) to the amazingly extensive possibilities of chip memory. The first known architectural drawing was found scratched on a pyramid; we had approached three- and four-dimensional renderings of an object. There was a degree of externalization beyond pencil and paper. Symbols could be manipulated in a mode notionally external to oneself.

The writings of Jean Piaget, Marshall McLuhan, and Jerome Bruner became pivotal. Bruner, president of the American Psychological Association, was one of Piaget's American champions. He was one of the American psychologists who had the courage to go beyond running rats through mazes. He helped make cognitive psychology respectable, and he refined many of Piaget's experiments. He confirmed Piaget's results but reached different conclusions.

Piaget's theory of psychological development was like metamorphosis. It posited a unitary mentality going from caterpillar to butterfly—through distinct stages where children are very different from adults and should be approached in a different way. Bruner's conclusion was more powerful. He acknowledged such stages, but he argued that instead of metamorphosis there were really separate mentalities. He spoke of the *inactive mentality*, a view of the kinesthetic. Piaget called it the *preoperational* way of thinking. Bruner's idea was another articulation of that revolution of the modern world in which children are seen not as undeveloped adults but as modally different creatures.

Bruner was well placed for a role in the post-Sputnik "curriculum reform" movement. His *Process of Education* was a summary of a 1960 conference concerning revamping the American curriculum. His *Towards a Theory of Instruction* (1964) is the best book ever written on how to approach the task of teaching—part because it draws upon theories of ignorance (a philosophical question since Ferrier, but rarely a cognitive one). It had long been perfectly acceptable to discourse on why we do not live up to our capacities, a subject Ferrier had addressed 150 years before. It is as if you wait until students are in college to teach them physics, and then you are surprised when 80 percent of them never understand it. By the mid-1960s, Bruner was designing curricula. He created a company in Cambridge, Massachusetts, and designed a curriculum called *Man, A Course of Study*. He took college-level anthropology and found ways to present it to fifth-graders in lieu of the usual subject of "social studies."

We all want to study what we find most interesting. And the chance of our being interested increases with the number of different ways we can address a problem. Most problems are resolvable along many different lines, and the computer helps us find the line that is most hospitable to our efforts and which helps to integrate the fruits of disparate lines of inquiry. We can find buried treasure by pouring water on the floor and seeing where it leaks down between the cracks. The computer can offer us such scanning, and can show us where we should dig.

Mathematics has been perhaps the greatest consecutive achievement of the human mind, but in the late 19th century theorists began demonstrating that our grasp of mathematical reasoning was misplaced. G. H. Hardy said "I am doing mathematics when I do not know what I'm doing."

It is an interesting experiment to get a 5-year-old, a 10-year-old, and a 15-year-old to try to program a circle. The 5-year-old inevitably wins, because his body computer happens to do the right kind of math. It is much better than the math that the visual computer does, and vastly better than the math from the 15-year-old's symbolic computer. The 5-year-old sees a circle as performative (what happens when he spins, operationally completed by a pencil) whereas the 15-year-old regards it as imitative (he pursues the idea of a circle). William Blake said we could be connoisseured out of our senses. Imperfectly assimilated axioms and doctrines bury elemental observation.

Two objects are important: selecting the targets that you are trying to reach, and then removing interference. A strange dialectic of creativity with difficulty awakens creative people who are engaged in creative tasks. Like the less persuasive model contrasting the left brain with the right brain, this asserts that "if we instrumentalize things one way, those engaged will reason that way; if another way, reasoning will flow down another channel." Or electroencephalograms can be taken over the life of a single person, to show which areas of the brain display activity at different stages. One's inquiring powers are built from a mosaic of subskills acquired over a long time.

The accompanying slogan is that "point of view is worth 80 IQ points." Mentalities are not likely to be uniform. First, this means that we should direct our teaching toward the mentalities that are best at learning. Second, it means that there is a strong chance of interference, that the mentalities are competing with one another in various ways, and, in particular, that the expressive mentalities are likely to compete better. Competition, said the classical economists, is disguised cooperation—and this is truer in the life of the mind than in most issues.

Can we develop a guide to learning, as opposed to randomly trying to use the computer or even to imitate Papert? One observation that emerged from Papert's work was that the children never created a tool in Logo. Children do not do tools in Logo. They create things that have various effects, but they do not sit down and build themselves tools. They can draw a picture in Logo using a turtle procedure, but they do not create anything like MacDraw or MacPaint that would give them a painting tool. This absence was troubling. One of the destinies of personal computing is to enable end users to create tools specific to their problems. The computer, after all, is one of those few media in which it is not physically harder to make a tool (defined as a technique, a grammar, or a set of rules) than it is to make something using a tool. In most aspects of life, creating tools requires special furnaces and machines. People would rather go to Sears.

But the notion of instrumentality is twofold. In the material world, a tool tends to be harder and more complex than what it makes. But in the information world, a tool is designed to simplify a complex universe by applying certain simplicities: it is a tool in the sense that a sieve is a tool. In fact, one of the earliest conscious tools is that for eliminating nonprime numbers, the "Sieve of Eratosthenes." Calculus is a tool, and one has to go further to obtain it than just to Sears. Tools function differently as interactive formalized systems or as reified objects of design. Tools are made in the human mind and transferred to the computer. The computer is not yet autonomous.

On the computer, however, one is using the same implement from which the tools are made. It is as though you could take a piece of clay and make ceramic tools that would improve upon the clay. Seeing the computer neglected as a workshop for tools struck Alan Kay as harshly as if workers in clay had themselves never thought of ceramic tools. To this end, some of the most exciting research since the early 1970s has tried to determine how to create an environment that would allow young children to see the possibilities at hand in the sensory world around them, as a small child in the forest reaches for sticks. Kay always wanted to simulate what this computer would actually do. And he wanted to try out these ideas on children, so it was natural that he focused on the work of Piaget and Bruner.

All the necessary awakening is deemed to be implicit in instruction. The question is how do we awaken, and the answer is radically lacking. Children are not precluded from making their own tools. But they are rarely sufficiently enheartened. The problem is attitudinal, and it is an extension of the "friendliness" question. Childhood may be a period of potential originality, but rarely is it a period of operational originality. There are only a few thousand jump-rope songs, although there have been billions of jump-roping childhoods.

Bruner argued that there were some teaching tricks one could play, besides the developmental stages discussed by Piaget. For example, children can be stimulated to think by making symbols into objects, moving them into a category that is more manipulatable by a younger child. Bruner also argued that there should be "body-level learning." A sound way of teaching physics, for example, is to get the student to experience physical phenomena, making the body the object in question, and then progress to seeing this experience as part of a continuum with the rest of the external world, not a special case. Finally, the student acquires enough physical intuition to be able to address the problem symbolically.

Basically, a good physicist's intuition of the shape of the answer is what guides him through the mathematics. Newton, for example, explained to Halley how a problem could be resolved by what we now think of as the theory of universal gravitation. Newton's intuition told him that the mathematics would follow. Halley was dumbfounded as Newton merely asked for a week to formulate the proof.

One challenge is to reconcile the "popcorn" state of a creative person with the calmer, organized planning that lies behind accomplishment. Insights into this are provided by the vast literature on perception and visual cognition. For example, if you randomly tune to a movie on your television it will take you about ninety frames—3 seconds' worth—to decide whether you have seen it before, even if you haven't seen it in years. What the visual system requires to prompt it correlates closely with a certain kind of creativity.

The visual system is highly permeable, for it is never at rest. The eye is always moving and lighting, in a scan rather than a stare. The symbolic mentality is exactly the opposite. It builds long chains of inferences and abstracts them. It needs to fasten on one problem. It would usually be fruitless to stare for 2 hours at the first thing you saw in the morning. But Leibniz might sit for days pursuing a chain of logic. Children try to solve problems this way, and as soon as they have an idea they commit to it so adamantly that it becomes a problem of its own, requiring that new data be fitted to it.

Most people who do mathematics concentrate for hours at a time. And that is one of the problems in teaching mathematics to children. We teach them what adult mathematicians talk about. Jacques Hadamard, in his monograph *On the Psychology of Invention in the Mathematical Field,* noted that around 90 percent of the mathematicians he had studied did not use symbolism in their problem solving. They used images or other figurative means, and many of them had actual body sensations. Einstein said that he had physical sensations that were kinesthetic or muscular. Many poets report experiences of synesthesia—a sight manifesting as a sound, or a color as a taste.

Such evidence correlates with other mentalities. We could conclude that schools are playing a cruel joke. We try to teach through the channels that adult professionals use to discuss what they do, but not through the channels that they actually use when putting themselves to work. This, of course, has been true for music. Formalist accounts of how things get done are extolled as models. There were no textbooks in fifth century B.C. on how to teach geometry. People taught because they were convinced that they had something important to teach. Tim Galway uses this "inner game" approach in his sports books, such as *Inner Tennis* and *Inner Skiing*.

This is all basically interference-removal theory. Children speak grammatically long before they know that there is such a thing as grammar. In fact, if one were not told about grammar, it is unlikely that one would try to invent it. Malay grammar, for example, is the construct of 19th-century colonial officials, but Malays have been producing complex and beautiful rule-governed speech for centuries.

People of achievement nearly always confabulate what they have done. Politicians are a fine example. They start from the end and they try to explain how they made that end come about. People usually do something from the bottom up and then explain it from the top down. In a sense, that is what we are doing in this chapter.

The purpose of studying psychology and teaching, however, is to achieve a user interface that really works. Consider Bruner's three mentalities: the kinesthetic, the visual, and the symbolic—remote powers with tenuous diplomatic relations. Yet we need to achieve synergy among these three mentalities. For example, calculus makes sense originally as the properties of a rotating cylinder or a descending particle and goes on to unlock the universe. Valuable work has been done by Oliver Sachs on what happens to people when they get brain lesions and have parts of their body map removed. It is apparently much more terrifying than being blinded; the victim has absolutely no reference points.

Research in the 1960s showed that people wanted to point at the screen. But what was needed was a world in which the computer appeared as an entity of humanity, and in which the user felt able to swim into the computer. The simplest way to develop such a body map is just to use a mouse or a stylus. Kinesthetic feedback lets you actually feel part of the computer's world. For example, a mouse was developed in which the amount of resistance depended on the size of what you were doing on the screen. If you were dragging something large, the mouse's resistance was strong. If you were working on something small, the resistance was much less. That little trick is amazing because suddenly your world on the computer becomes more real and vivid because of kinesthetics. Thus the Scots commonsense philosophers argued that the prime proof of an external world was that we felt it resisting us.

The mind is also very good at recognizing images, and at scanning, so we want something that is as much like a bulletin board as possible. One can recognize a certain icon out of a hundred about 4 times faster than one can find a certain word in a list of a hundred; a different part of the brain is at work. So we want a larger screen that is image-based rather than text-based. The iconic, multi-layered mentality is also good at making comparisons. A good icon prompts you in many ways. Much of its inference comes from juxtaposition. A cigarette company would never dare claim that smoking makes men handsome, but by having its advertisements feature some unspecific handsome men it can get your visual system to make that inference. It is a weaker inference than would be made through symbols, but it is made nonetheless—and in a way that bypasses the symbolic mentality's ability to dismiss the proposition as rubbish.

In view of these two mentalities, it was clear what to do. Multiple windows and icons evoke the visual dimension, and the user is able to move them kinesthetically as though they were real. It is more difficult to energize the symbolic mentality—to put yourself at its disposal. That was why an object-oriented language called SmallTalk, a symbolic discourse about all the fake objects on the screen, was invented. The whole approach to working with children, therefore, is to get them to do things physically, kinesthetically, and iconically while doing most of their reasoning in that space, and then write down the results of their reasoning symbolically.

People have a very high internalization capacity, and there are innumerable media of communication. The ability to handle fiction, for example, is a uniquely human capacity conditioned by culture, education, and intelligence. People develop the ability to become engaged with something that they know is not true, not realistic, and so on. This shows the richness of possibility. The degree of engagement that is possible is a reminder that many dimensions of human capacity remain unmobilized.

McLuhan's statement that "the medium is the message" can be interpreted to mean that, in order to receive the message engraved in some medium, you have to cross over into the medium (much as a radio must synthesize the carrier frequency in order to subtract it, leaving the signal). It is unfortunate that McLuhan did not offer a clearer elaboration of one of the major insights of the 20th century. As he put it aphoristically, "I do not know who discovered water, but it was not a fish." Yet function is likely to precede consciousness.

Which comes first, the environment or the ability to abstract it? In a real sense, the fish's DNA and RNA discovered water. There is a steady movement toward form. So in order to get somebody to do something, you need to create or to fit yourself into an environment. In order to get messages from the environment, you have to internalize it.

For instance, millions of people now use Apple Computer's Macintosh. Whether they want to or not, the Macintosh gently gets them to intensify their kinesthetic and symbolic mentalities—regardless of whether they are using spreadsheets or MacPaint. All interaction is likely to involve inducing this in different intensities and proportions. The interface is the same from application to application. What the user internalizes is not the different applications but what he needs to do in order to get the applications to perform. This is an interactive relationship.

This is one reason Macintoshes are used more hours per day than IBM PCs in the *Fortune* 100 companies. The Macintosh is less fatiguing because the user is constantly being exposed to different ways of thinking. He is running, not pounding in lockstep with the metal monster. Other designs pull you along instead of allowing you to have a more complex, mediated reaction involving the body with real possibilities of alternating approaches. Of course, this is an idealization; there can be no patient interaction until far more is known about humanity.

The importance of these theories is not whether Bruner was right or wrong; his main objective was to stimulate. His work formed a solid metaphorical basis for creating designs, and he never claimed that his theory was cosmic truth. But a theory that helps you generate other ideas is worthwhile. The real, functional objective of science, after all, is not to obtain all data but to raise the level of analysis. Gods may absorb a universe without mediation; mortals live in ideas and generalities, filtering an intolerable body of raw givens.

The literature of educational psychology in the United States has generally been appalling. It is as if there were a conspiracy not to study accomplished children. For example, there is minimal discussion of how well a child might be able to read. America's exaggerated attempt to be democractic and balanced leads to a reluctance to even talk about how and why some children perform better than others. There is a passion for standardization. It is assumed that it is the system's business to bring a model citizen into being—and to see his alternatives as rivals.

We certainly have little faith in fixed ability. The variability in any given child's responses, and in what people of the same family are capable of at life's different stages, contradicts the notion of a fixed capital of innate capacity. Living multiplies what life confers. Yet America also has a romantic, individualist culture that believes in inborn gifts—that individuals are good at some things and not others. Demands for intensive effort at things which are not enjoyed are regarded as coercive and intrusive.

One needs to look at a child's performance in maximizing interaction—voluntary sports or music rather than the slog of the classroom—to explore, first, the method used by the teacher and then, second, how the child could become as accomplished as the teacher. Moreover, we need to examine which teachers can communicate their capacity and which cannot—and why. Suzuki, the Japanese violin teacher, has been a huge influence on Alan Kay's work.

We agree with Suzuki that whatever variation there may be in ability, or however many standard deviations one is from the norm of what is to be accomplished, is generally irrelevant. Suzuki assumes that higher capacity is possible when there is an effective relationship between teacher and pupil and when society has high expectations rather than high accommodation. Susuki induces the proper attitude and legitimizes intense demand—reciprocal demand, for the pupil must ask of the teacher as much as the teacher asks of the pupil.

In any event, one does not know one's ability until it has been halfway elicited, which hardly ever happens. This impression about individual potential is corroborated by Glen Doman, who spent many years repatterning brain-damaged children and who discovered that they could put to work other areas of the brain to compensate for the areas lost. He then began experimenting on children who had not been brain-damaged. The evidence of individual potential that emerged from his Better Baby Institute was astounding.

Doman's work was very similar to Suzuki's efforts to repattern brain-damaged children by teaching them the violin. Suzuki believed that talent is what you can exercise with little practice, and that it is manifested in innumerable different combinations. But he also believed that through practice we can accomplish most things that interest us. The level of interest usually determines the level of practice. Learning to play the violin, just like learning one's native language, involves innate capacity, age of exposure, degree of confidence, and the like. Suzuki argued that talent shows only in the top few percent. Anyone can learn to pole-vault, for example, but the bell curve will still show itself—there will be a few Olympians and a few klutzes.

It is also worth noting that there are many forms of practice. It can be proactive, such as learning to vault over higher and higher obstacles, or it can be passive, such as listening to someone else until you can express yourself. All achievement is open-ended, and it is very much a question of the standards we set for ourselves. Probably 90 percent of Japan's so-called economic miracle can be credited to that country's excellent high schools. The universities are notoriously poor, but through good secondary education Japan has mobilized 100 million competent people.

However, there is a difficult question behind the task of re-creating an environment such as Athens: Why do some societies distill only minute dewdrops of exceptional talent whereas others produce a flood? Papert believed that the right computer environment, all by itself, would be enough to move us toward the creativity of Athens. But it was noticed that the children working with Logo were not creating tools. Was that because children simply do not create tools, or because they were not in the right environment? Using SmallTalk, countless children have created tools.

This approaches the key question. Bruner said when he was interviewed at age 70: "If I had to do it all over again . . . I would do lots more peer teaching with the kids because they are frightened of the teachers." We can actually move toward building an environment with a computer—an environment where one's growth in imagination and understanding grows retrievably in the machine as well as in one's mind, whereas the world elicited by Tolkien or Darwin must be retrieved by mental effort each time. We can interact directly with the views of McLuhan and Bruner. Feedback and synergy are attained.

Man's primary environment, however, is the mind. A persuasive book can indeed create an environment even though it is not formally established. We are talking about two different ecosystems. Two people playing chess on an imaginary board have created an environment. But the computer has vastly greater possibilities. Although it is the purpose of the human mind to create environments all the time, the computer can make environments more elaborate, more determinable, more enduring, and more nearly communicable in real time.

### The Athens Interface

Comparing the invention of the printing press to the arrival of computers has become a cliché. The printing press transformed society, but selectively. More stayed the same. By the same token, the computer has not obviated nuclear weapons, just made them more accurate—if perhaps more controllable. We may have come to think about some things differently. But there has certainly been no moral impact from this diffusion of knowledge. An increase in the sense of possibility is not a transfiguration of vision. People see deeper into things than into themselves. They kill one another with greater gusto and more efficiency than ever.

To be sure, the general level of education is higher than when people had to study stained-glass windows to get a picture of the world. But something has also been lost in understanding that world. There was a far greater physical interaction five centuries ago. In the 21st century, an astoundingly small number of people will take advantage of the many ways that will be available to interact with, let alone to understand, their world.

McLuhan observed that one was educated in Oxford in 1200 A.D. by going to a room where thirty people sat and listened to someone talking. Notes were taken. After accumulating volumes of notes, a student would receive his robes as an educated person. McLuhan's point was that little had changed. But he exaggerated the passivity. He underestimated the degree of argument. Moreover, people had to use their memory much more, and not always by rote. They lived with Aristotle, Thomas Aquinas, Duns Scotus, and Boethius in their heads, not on their desks.

By increasing dependence on books rather than on memory, the printing press increased passivity. The printing press, however, vastly increased one's ability to get a second opinion. Only the greatest universities had rival professors, and only a few students might interact with them.

The printing press served multiplicity. And if technology is good for anything, it is extending choice and range. Yet only the person who wants a second opinion is likely to get one. The instrument is passive, and the problem remains attitudinal. It is therefore unclear how—or whether—the computer will help educate the individual. Some of the best libraries in the world, after all, exist within half a mile of America's most devastated inner cities. There is no stream of hungry readers searching for anything, let alone a second opinion.

The computer, however, is understandably more interesting to children than are dusty volumes in the New York Public Library. It is more like television. But television is an exceedingly bad medium for learning how to think. Yet many educators who develop curricula ensure that computer use is ever more like watching television—and, specifically, television commercials. Neal Postman, author of *Teaching as a Subversive Activity* and *Teaching as a Conserving Activity,* warns that when you see Carl Sagan on television lecturing about the cosmos, you are concerned more with Sagan's haircut than with astrophysics.

Television is at its most dangerous when it is trying to be cultural. And this might be because it is trying to seek a certain back-slapping accessibility. Also, the computer's very facility is the reverse of a time binder—you can stop and come back, come back and go away. If there is virtue to consecutive concentration, letting the fish swim great distances in the sea of the mind, it may suffer from an instrument that makes thought bites as easy as sound bites.

The means of enhancing education are not going to be technological. The Open School in Los Angeles, with its ethnic and racial reflection of the city, is a pivotal illustration. It is an excellent public school in which there is no preference for IQ and in which many of the students are Hispanic. Busing has made it a demographer's paradise. The principal is not interested in examining math or science or art or music. She focuses on what it means to teach children how to be proactive about their own learning processes. The school is like an Outward Bound program in the jungle of knowledge. It is established on Brunerian grounds. Alan Kay is using this school to study how children learn.

What is so riveting about the computer is that it will have all the transforming influence of a book, and more. The computer, on the average, has a better chance to influence than does a book, on average, because the computer can draw the user in. The best books represent a world already made, because they represent concentrations of talent. They present powerful, reasoned, and non-real-time experience. With the computer, there is unprecedented synergy and feedback between the student and the medium. The purpose is to adapt to what one cannot imagine. Open-endedness is crucial. However, there are times when synergy is not desirable—when the student needs to be cornered and forced to overcome obstacles.

We do not want to put Athens in a computer. Instead, we want to use the computer to bring forth the next (and, we hope, greater) Athens. The computer is a ball-and-wall encouraging children's use and developing their legs and coordination. Where they walk and climb is the next decision, not identical with learning how to.

Today, most schools are trying to determine what to do with their computers. Parents and school boards tell teachers to use them. And often they are used. In Los Angeles, many a classroom has a couple of Apple II PCs. But having two computers in a classroom is like having two pencils there. Having a "computer lab" in a school is like having a special room with thirty pencils that the students are allowed to visit twice a week. The computer is not taken for granted. Few things can be used in a first-class way when they are not second nature. If one has to be painfully conscious of how to use something, it won't be used particularly well. So any results of how children learn with computers are essentially meaningless.

Each child must live with a computer as literate people live with writing materials. Otherwise we cannot judge the effects of the computer on education. The child has to use the computer not just in the classroom but for mundane things. It is like the use of language. You simply cannot have a "language lab" where thirty people are allowed to talk for an hour a week and then expect them to acquire the immense repertoire of sentence or syntactic capacity that defines a language user.

After visiting Germany, Suzuki concluded that German was hard to learn. But he saw the ease with which German children learned the language. He then wondered what it would have been like to have learned to play a musical instrument in the same way. An hour or so a week of music practice hardly created an environment similar to the home life of a child in Düsseldorf. We agree with Suzuki. We are searching for what he calls "the mother tongue method."

The Open School has countless visitors. They all exclaim over the children's use of computers and say it is the best school they've ever seen. But this school was already an excellent one before the computers were introduced. The computers are integrated so well in the school precisely because the school doesn't need them to engage the children in the enthusiasm of learning. The students could benefit just as much from having Archimedes' sandbox in the classroom. However, they can take advantage of several added dimensions from the computer. But they are not yet able to exploit all the capabilities that the computer will offer. Not merely does the computer have to evolve, but so does our imagination with regard to its use. When we try to envision the role of the computer in twenty years, we see two consequences for education. There is the "probable" impact and the "could be" impact, much as there would be if we asked what influence books have upon learning. Consider how books could be used if we really wanted to use them effectively for teaching. The results could be fantastic. The technology is certainly profound: 2.5 megabytes, solid state, priced at a few dollars a megabyte. And there are 100 million different selections to choose from. Yet the invention of the book index is barely 100 years old. The organization of books is a continuing technology, always huddling behind the capacity to produce books. People need to see books that are efficiently organized. The point about books, however, is that they are not being used effectively at all. There is an excellent chance that the use of computer technology will be similarly poor and ineffective. Most talent is wasted, and most available techniques are misapplied or misconceived.

One danger ahead is that it will be decided that bumper-sticker education is enough for everybody. Two or three generations of television will have passed. All learning might be done through the computer, but it might be done on the passive level of television consumption. Education might not be undertaken in a way that requires deep thought or reflection.

Hypercard illustrates the pitfalls. Built into Hypercard is an array of special effects, which children love. Each transition is made interesting. One step dissolves into another. But much of the information that one might receive from Hypercard is offered in the form of "bullets"—essentially, bumper stickers.

Confidence needs to be developed at an early age. Broadly speaking, virtually all education is passive in comparison with what could be accomplished. For example, professors of English who have been fed on the cream of their language usually write very poorly. So there may be a point in instilling some wariness into the computer. The user might be penalized for giving the rote answer. The computer should not just charm people along, as in a good expository discussion. There is a limit to what charm can accomplish. In effect, this is what is meant by Socratic demandingness.

The search for what we might call "the Athens interface" continues. Let's suppose that a computer scientist with the brilliance of Newton could somehow create the environment of Athens. This would be far more significant than the invention of printing. The printing press may have nurtured the Renaissance, but that era did not approximate Athens. The Renaissance was an overwhelming experience rather than an empowering one. There was greatness, but the Renaissance never produced an Aristotle. The crucial question remains this: How do you build Socratic demandingness into a computer? (Or, to be blunt: How do you get the computer to bite you in the ankle?) There is no need to care how well a computer plays chess if it does not feel bad when it is beaten.

Let us return to the metaphor of television. Television has convinced scores of millions of Americans to tolerate an impoverished dramatic form. It has lowered people's expectations to a level of formula. The difference between television and the computer is that the latter is fundamentally more productive. Arguably, television is not blameless in eroding the family, although its penchant for formula can also reinforce traditional forms (the hero rarely gets killed, and justice usually triumphs). One concern is that inadequate attempts to achieve the Athens interface with a computer might result in something with television's increativity and sedativeness.

About 40 percent of the efforts in the Open School are spent on "anti-television"—simply combatting the passive state of mind that television induces in children. Children's opinions of Socrates could easily be lowered drastically. They would cheerfully accept a two-bit Socrates along with a two-bit Athens if there were enough flashing lights and whistles. Would they tune in to anything else? Again, we would be stuck with bumper stickers.

Television has indeed given children access to a larger vocabulary, albeit passively. But it has not given them access to the ability to deal with the syntax of complex ideas. Learning is enhanced by argument, by trying out ideas on friends. On one hand, so much has been accomplished in the last half-century. There has been enormous access to an immense range of possibilities. Sharecroppers' children have gone to Harvard, and children of Harvard alumni have become craftsmen. On the other hand, so much of what once excited curiosity has been whittled away. Television's greatest crime is habituation. Years ago, the family doctor was seen as a scientist, the family lawyer as a jurist, and the local businessman as an expert in management. There could be a sense that one's environment was intensely hospitable to intellectual activity. That feeling has evaporated. Instead, one sees Lee Iacocca, Carl Sagan, or even some Nobel laureate on television. These are the mountain peaks which one can see on the tube but with which one can hardly interact. Meanwhile, the nursery slopes have been dynamited away. There is the need to restore that direct, stimulating environment—an environment in which questions can be asked and answers given, as in the Open School.

What is needed at the very least within ten years is a sequential, logical scolium in which the student is called upon to draw the next conclusion, to formulate the next collage. It would be awful to abstract, from the knowledge of distant historical action, a serious program that compelled one to make the generalizations of classical mechanics that we call the relativistic syntheses. For example, Henri Poincaré was groping in the dark when he said that the next stage in physics was to produce a doctrine in which the limiting factor was the speed of light. He did not know that Einstein was at that moment bent over a desk, scribbling frantically. The easy stopping places must be avoided.

We will all be swimming upstream as long as television—and the classroom technologies redolent of television—is the main informational environment. Puzzlement, however, is one of the great pleasures of life. It is the parent of wonder. The Open School, for example, tries to restore puzzlement to many of the children who had been soothed and subdued by television. Television has given us a sequentially various but attenuated environment, with simple descriptions and with a small number of categories for the entire world. That was why McLuhan called it a cool medium, and why its teaching, as he said, is numbing.

Yet we are optimistic about what could be accomplished in education. The example of Jaime Escalante teaching calculus in East Los Angeles is as convincing as it is popular. Escalante used every trick in the book, including shame, and played every sociological tune that needed to be played in teaching advanced calculus to his barrio teenagers. They were molded into adopting his set of goals rather than the ones they had brought from the streets. We need to build an Escalante into the computer.

All stages of human development are as much shaped by input as by output. But advanced calculus is not more complex as an abstract operation than good basketball is as an externally performative one. If it were, only a Newton, a Pascal, or a Leibniz could do it.

The best that could ever be done for education would be to design something that children from the ages of 3 to 6 could use before they started school—something that would enable them to have their questions answered without necessarily depending on the presence of adults (although adults could help the child to frame his questions). Since children at this age can handle computers, they might also be expected to read. Yet after working with vast numbers of children, Suzuki concluded that there is a variation of at least a year and a half in the rates of their development, and that this is already apparent at a very early age. This is not genius and retardation, just a standard distribution of complex capacities and developments.

There are several parallel ways of learning words, symbols, and the like. A small child using a computer might know the command **save** in the same way that she would know a Chinese character (although the command is specific to the computer and the Chinese character could appear anywhere). Most American children can identify a McDonald's by the golden arches well before they learn to spell the name of the company. A child is privy to what seems to the child a large number of arbitrary signs.

It is fair to ask whether government shouldn't assure everyone access to a public network, much as everyone has access to a public library. Free credits which could be used every week might be given to information banks. Along the way, computers will appear in the home much as pocket calculators did.

In the late 1960s, Dartmouth opened its time-sharing system to everyone in Hanover, New Hampshire, for public computation. But the approach did not acquire momentum. John Kemeny, the computer scientist who was president of Dartmouth, said that he just wanted to get people's conversation up to the level of the pre-television age.

We would now like to see a widely diffused user interface in which the computer might conclude that it will stay up all night, while its owner sleeps, in order to use some of the free credits to examine material to see if anything correlates with what the owner has been doing all day. The owner would be presented with the results in the morning. Something similar, called Newspeak, was built at MIT in the early 1980s. It interrogated news sources all night and emerged with an idiosyncratic newspaper in the morning.

There is a difference, however, between having an agent and having a tool. Newspeak was like using an encyclopedia. The cross-references in an encyclopedia represent the scope of interest of the editor and the contributors, rather than that of the reader. The denouement of a detective story—where Holmes or Wolfe is alerted by a child's chant or by parsley sunk into butter—is idiognosis, particular insight, not available from reasonable expectation; yet the user can be seduced into distraction. It is easy to be drawn into spending hours in a complete text-retrieval system such as NEXUS, even if one is searching for something simple. This is what also happens with Hypertext. Yet the fringe material often proves to be the baseline. One's strategy of inquiry changes. And this is the value of Mac and of Hypercard.

No tool is going to enable us to handle the material in the vast information base of the future. Instead, intelligent and autonomous processes must work on our goals. We will have a continuous loop in which the process will keep modifying the goal. This can be called "goal cloning." We need to clone our goals into processes and let the processes (called *agents*) constantly search the network for us. But these agents tend to be anthropomorphized—at least at the level at which we anthropomorphize pets. Disappointment results. It will be years before an agent has the human flexibility which we now need. Thus the great autonomous molecules of the primeval sea became in due course parts of cells, and these cells became instruments of larger organisms—transmissable information but set to use in ever-more-complex undertakings.

The problem is similar to having a new household servant from some remote culture. The dialogue will be amusing, initially. We do not expect our agent to try to conduct the equivalent of a Socratic dialogue. But Socrates encouraged learning by making it challenging, and at times enraging. What Hypercard tries to develop, therefore, is an environment that is easily summoned to examine things about which one is skeptical—rather than depending upon the perspective of the servant.

McLuhan claimed that the book introduced "original man." The computer can create "skeptical man," because it lets you interrogate so much more easily and widely. There are more ways of conducting experiments and especially of getting alternative views. This last contribution is crucial for learning. Marvin Minsky said it very well: You do not understand something unless you understand it more than one way. That is the primary goal of the Open School: not just to give the children the understanding that there are multiple viewpoints, but to give them a sense of which viewpoint is going to do them the most good in a given situation.

The user interface can expand by providing a more powerful fringe environment, to make the user tack against the winds of inquiry in order to broaden his approach. This allows the user to observe that something doesn't look right and to deal with trivially set up simulation.

By mid-1990, Seymour Papert had accepted a $3 million grant from the Japanese game company Nintendo to continue studying the relationship between play and learning in children. He began exploring the unmet psychological needs behind a child's fascination with such hits as "Super Mario Land." Children, who are generally infantalized in American society, escape to an alternate reality with Nintendo. They turn to other worlds when they are prevented from confronting important issues in their own. "Schools dictate a particular way of thinking," Papert says. "If you happen to fit the mold, the experience is affirming, but if you don't, it is a constant putdown."

Schooling today is somewhat simple, and perhaps therefore somewhat appalling. More and more technology is constantly being invented. And that can provide far more access to information. But such an energetic arrival of technology can also make it easy for people to ignore the creative possibilities and to become dynamically passive, using the obvious to get the extraordinary rather than discerning the less obvious to achieve the miraculous. The 20th century is the century in which people have had more access to more things than at any time in history, and with no comparable ability to use them as they arrive. The great problem of the proliferation of information is that it produces such a vast quantity of material that it can drown one's sense of confidence, leading to the feeling that "the system knows more than I do." The future of our possibilities is not how well we learn, but how well we unlearn what is inappropriate or based on false processes.

### Acknowledgments

The authors want to acknowledge the insights and information from Alan Kay of Apple Computer, Inc., an Apple Fellow who is best known as the pioneer of personal computing. We are grateful for the insights he shared with us. They were vital to shaping this chapter.

# 11

## Touching Everywhere: Intensifying the Human Environment

*Lee W. Hoevel*

Ask a sociologist to define "human interface" and he might respond, "It's a symbiotic relationship between man and machine." To the linguist, it is the translator between computer talk and people talk. To an office worker, it is the display screen, the keyboard, the mouse, and maybe the fax. To the computer scientist, it is "the layer of software between the computing system and the human user that interprets the user's input into low-level system language that the computing system can recognize and act on and that interprets output from the computing system into a presentation to the user through which he can gain feedback and get more information."

The preceding paragraph illustrates two points: First, the term "human interface" means different things to different people, and, second, *never* ask a computer whiz for a definition.

We have all had experiences with human-to-machine interfaces, whether or not we recognized them as such. For many of us, the first major opportunity to see how man and machine interact began when we learned to drive a car. High school vocabularies suddenly expanded to include words such as *throttle* and *accelerator. Choke* and *clutch* became nouns instead of verbs, and they somehow crept into our everyday conversation. Those who learned to drive before power steering became common grew used to wrestling with balky steering wheels as they maneuvered into tight parking spaces. The bonding between driver and machine—the human interface—was clumsy at first, taking many months to master. It was some time before we could drive smoothly and effortlessly, before we could shift our attention from how to operate a vehicle to simply using it to get where we wanted to go.

Innovative and effective human interfaces now pervade our personal lives. From the automobile that tells us when a door is ajar or when service is needed to the oven that stops cooking when the roast is done and the automated teller machine that dispenses money at our convenience, interfaces are performing two major tasks: they are making it a pleasant experience to use a variety of machines as helpmates, and they are protecting us from our own ineptitude. (A modern automobile's microprocessor, for example, would probably have prevented us from flooding the engine or locking the wheels of our first cars.)

The notion of a friendly and useful human interface is moving into our professional lives as well. The movement is being pulled by our need to pick from the flood of information rushing past us and pushed by the wholesale delivery of processing power to the individual. We are bombarded with undigested bits of data, most of which whirl by in a twinkling without being integrated into useful information.

In this age of information and knowledge, news of change is delivered so rapidly that we have trouble sorting out what is significant. It has been hypothesized that we receive 1000 times as much information in a year as our grandparents did in a lifetime. That is an information-explosion factor of 10,000 every decade. We can only wonder how our grandparents made do without our information pipeline and how our children will survive with it. Microprocessors such as Intel's 80486 may well become "information firehoses," spraying numbers, facts, graphs, words, and images toward the user. The human interface must provide a method by which man and computing machine are bonded seamlessly, working as one to capture and use information.

### The Challenge

In high-security laboratories around the United States, private researchers and the military are struggling to develop remotely piloted aircraft. The technical implications of this effort are staggering. Some of the most advanced applications of human-interface technology are involved. The pilot's steering mechanisms must be extended past the physical boundaries of the airplane, so that the machine can be controlled remotely with the same look and feel as if the pilot were seated in the cockpit. The data glove, one of the newer HI devices now being tested, allows the user to manipulate images of knobs, buttons, levers, and the like to initiate the same actions that would result from physically touching the actual controls these images represent. Advanced technologies applied to military applications eventually trickle into the private sector and find various uses in business.

As these experiments continue, noncontrolled (as opposed to uncontrolled) experimentation with some of the same technologies is going on in homes across America. Video games are pulling their players into a make-believe environment in which machine and operator act and interact.

Joysticks, control pads, and now the data glove excite and enlist our senses while extending them to the machine. The challenge in the field of computing equipment is to secure the same level of interaction and bonding of man and machine that we see in pilotless airplanes and video games. But to date HI (as applied to personal computers, which will be the HI target during the development wave) has not benefited from the funding and the focused research of the military R&D programs or from the unbounded imagination of the video game developers.

### The History of the Problem

In 1971 the Xerox Corporation set up its Palo Alto Research Center (PARC), which soon attracted a pack of talented computer scientists. The alliance between these scientists and Xerox resulted in the prototype of a new kind of computer called a "personal workstation." Xerox developed the Alto, a computer intended for the use of an individual.

The Alto had significant processing power and memory, a high-resolution bit-mapped display, a keyboard, and a pointing device called a *mouse* (originally developed by Douglas Engelbart at the Stanford Research Institute). The inclusion of the mouse may well have resulted from experimentation at PARC in which Stuart Card, Tom Moran, and Allen Newell compared four text-selection devices: the mouse, a rate-controlled isometric joystick, step (cursor) keys, and text keys. The mouse was judged clearly superior for text selection by the criteria of positioning time and error rate. The same group also pioneered the description of the human information-processing system known as the Model Human Processor, which focuses on our capabilities and limitations as processors of information.

This was one of the first attempts to model human capabilities and to describe expected human interaction in terms of the human perceptual, motor, and cognitive systems.

The Alto was the first link in a chain of HI development although it was not so recognized or marketed. Alto's first descendant was the Xerox Star, which focused on a new interactive technique that could fully use the capabilities of the personal workstation. The Star seemed to be the first system to employ pattern-manipulation capabilities directly. Even in the absence of commercial success, the Star left its mark on HI by encompassing many of the user-interface ideas being developed at PARC. These included:

a conceptual model familiar to the user

seeing and pointing, rather than remembering and typing

"what you see is what you get" (WYSIWYG)

universal commands (e.g., MOVE, COPY, DELETE) that were the original commands incorporated into the Star interface

consistency (e.g., if the left mouse button is used to select a character, the same button should be used to select a graphic line or icon)

modeless instruction (e.g., one could select an object and then invoke a command or an action, without having to use a particular mode to access an object)

user tailorability (e.g., moving windows around a screen, defining aliases).

The Star was never a commercial success, but it inspired other system designers to build interactive computer environments around the user. Apple seized the opportunity to capitalize on the Star concepts and produced the Lisa in 1983. At a price of about $10,000, the Lisa did not overwhelm the market. It did, however, give rise to the major success story of direct-manipulation interface workstations: the Apple Macintosh.

The direct-manipulation style of interaction that characterized the Star, the Lisa, and the Macintosh was quickly popularized by video games. Its central ideas include visibility of objects and actions, the ability of the operator to point and click, and rapidly reversible incremental actions—all of which lead to feelings of competence and mastery of the system by the user. The basic form of a modern HI had been established.

## The Evolution of HI Orientation

HI, by definition, covers all aspects of human-machine inter-action and is modeled in three layers—the physical, the opera-tional, and the cognitive—that apply equally to humans and to machines. Early human interfaces of the 1960s and 1970s es-tablished basic interaction through rigid structure. Man and machine were physically bound through predictable commu-nication protocols, with keyboard character choice reflected by confirmation of keystrokes on the screen. This relationship changed dramatically with the advent of the IBM PC and the Apple Macintosh.

Apple probably affected the human interface more than any other technology or market force. Its influence moved HI from a rigid physical union to a more flexible, operational one. At the same time, computing power moved from the computer room to the desktop.

Users quickly adapted to the more dynamic and flexible style of meaningful communications. Operational concepts like "windows" and "mousing" enlisted legions of new users. HI be-gan to change the notion of "computer" from a complicated power center accessible only by experts to a personal helpmate.

The next stage of this evolution will move HI from an oper-ational to a cognitive orientation—a orientation in which we will expect to communicate with machines as we do with other hu-mans. Next-generation HI will make sophisticated technologies available to the user in a transparent manner, so that no special training or knowledge is required. If we have to use manuals to operate our information machines, then HI will not be doing its job. Indeed, the unskilled user will stylize a personal method of communicating with the system, expecting (and eventually demanding) an environment where miscommunication does not occur.

The human interface is now envisioned as the gatekeeper for (or the window to) the effective use of computing resources. This was not always the case, for its job used to be to dissuade unskilled users from becoming directly involved. HI's new role is to be seductive, attracting and helping as many users as pos-sible. As the human interface has evolved in its role with the user, it also has undergone a significant technical change in its relationship to other components of computer systems.

## HI's Evolution among the Systems

Before 1965, there was little distinction between application, human interface, and systems software. During the following decades, operating systems became an important and separate component of the overall system, and were reused from system to system. During this period the user interface remained embedded in application software, although some graphics utilities and interfaces did emerge.

There has been a trend since 1985 to strip HI utilities and drivers from application and systems software, repackaging these functions as portable HI toolkits. The next step is to standardize the interfaces to these toolkits, making them an acceptable layer of systems software. It is important to note that some popular user interfaces, such as X-Windows and Presentation Manager, are being ported to many system environments, making them perhaps even more reusable than the underlying operating systems.

The human interface will continue to be the most visible and important transferrable part of a system. HI components will operate on a variety of hosts, without regard to hardware or operating system, and will be sensitive to the demands of application software. A system's HI is synonymous, in the eyes of users, with the system's accessibility and computing power. HI is of critical importance to vendors and users alike, affecting visibility in the marketplace, reusability, and applicability across a range of systems. Recognition of this importance is just now becoming obvious through vendors' commitments of R&D resources to advancing the state of the HI art, as well as through their marketing efforts.

## HI Presentation Methodologies

### Near-Term Orientation

In recent years, there has been a trend toward the use of icons, pointers (such as the mouse), and spatial and sound mechanisms in general to stimulate multiple human senses in a variety of ways. This encourages the full involvement of the human with the machine by simulating the ways in which humans communicate with one another.

The human interfaces on personal computers enable humans to solve problems in the context of space. The power of a graphical desktop is that it creates the space within which we can do our jobs better and more naturally.[1] For example, I might call for the blue file folder on the right side of my desk. That is a spatial way of asking for information, and computers like the Macintosh handle such requests the same way that a human would in a natural setting. The Mac, for instance, allows the user to delete a file from the desktop by dragging it across the visual display screen with a mouse until the image reaches a picture of a trash can.

In the dimension of sound, synthesized voice patterns generated by the computer (such as those found in NCR's talking automatic tellers and retail checkout systems) will become routine as output, and voice/speech input will become more common. At first voice recognition will be available in simple applications that require limited vocabularies, but as recognition techniques evolve more sophisticated systems will appear. We can expect voice annotation and dictation of documents, and the use of generated sound with video-disk sequences, to be intermixed with normal computer output sequences.[2]

The use of pointing devices, such as the mouse, will become more popular, although each user will find some kind of balance between pointing with the mouse and indicating actions with the keyboard or the voice. Some additional way will be found to point to the screen without taking the hands off the keyboard, such as using a foot to control a mouse-like device. Another approach, the thumb-activated cursor control located under the space bar, might prove successful.

Graphics will continue to improve,[3] and the use of icons will become even more popular. The use of screens full of menus will decline in favor of icon-oriented selection trees. Powerful functions will be buried in the operating system and will be available by pointing to an icon, or by moving an icon from one place on the screen to another. Several of these notions have already been embodied, beginning with Xerox and Apple icon paradigms, and now are incorporated into the electronic page paradigm of the Wang Freestyle interface.

Much work remains to be done to understand the human factors inherent in such developments.[4] For the first time, there are job descriptions for human-interface developers, and HI development teams include people representing the spectrum of disciplines involved in understanding and mimicking human-to-human behavior in a human-to-machine environment.

Computer engineers, software engineers, AI gurus, graphic artists, industrial designers, sociologists, linguists, psychologists, and supermarket clerks are together developing the HI of the future.

### A Vision Of The Future

Developers are working to provide virtual power at the desktop. That is, the total computing power of an organizational unit such as a department will be made available to an individual user through a distributed client/server approach. The human interface in this case allows the user to visualize his desktop computer as part of an integrated system in which there are many powerful servers. These servers are not necessarily in the desktop system itself, but they are part of the family of computers whose resources are available to provide services for the user. They will constitute a single-image model in the user's mind.

The process by which an action (e.g., getting a file or printing a report) is invoked will be uniformly understood across all the servers, regardless of their locations, and will be addressable from any authorized desktop. Thus, files may be stored in a distant database server, but they will appear as files or documents in the local desktop system. The "single-system image" will permeate the entire distributed system available to the user over an invisible network. Group resources will be available to the individual user. The human interface will see to it that the user's request is directed to the proper server(s).

The future will also bring human-to-machine communication that mimics the way humans interact with one another. Machines and humans will talk to one another. Machines will watch humans for body language that suggests actions the computer should take. For instance, a waitress may signal "two eggs, over easy" through hand gestures. Tomorrow's human interface will be able to train itself to recognize users' gestures that are intended to trigger actions—in this case, telling the cook to plop two eggs on the grill. In the office, gestures may be used to signal the conclusion of a meeting or to refuse an incoming phone call.

The interface between man and machine will be seamless—that is, it will be natural and flowing. It will be difficult to tell where man stops and machine takes over. For instance, the human interface on a supermarket checkout terminal will eventually allow the machine to recognize and verify a customer's voice or handwriting, and to authorize a check or an electronic transfer of funds without the involvement of a checkout clerk. A desktop interface will turn itself on when the user first speaks or gestures in the morning, and will perform routine house-keeping functions such as checking the electronic mail box and the voice-message box on the user's behalf. The HI will prob-ably also brew the morning's first cup of coffee to perfection!

The strategy for meeting the needs of end users is to prevent miscommunication between human and machine by having the system recognize the user and by providing expert advisors that interpret and guide the user's interaction with the system. There are three thrusts to this strategy.

The first is a thrust toward traditional information technol-ogies. In addition to providing "factory built-in" application-specific knowledge, learning technologies will also be applied selectively to increase the flexibility and life span of future com-puter products.

The second thrust is to expand bandwidth—essentially, the diameter of the pipe through which information flows. The purpose is to deliver maximal power and bandwidth to the user through the use of advanced power-boosting technologies. Re-sources should be used effectively and efficiently, so that nei-ther the user nor the system is saturated by HI requirements.

The third thrust is to make the user feel in command of the system without having to become an expert at using computers. This harnesses the power of reasoning systems and other forms of semantic modeling that enable diagnostic, error-recovery, and user-cuing procedures to maintain a positive work flow, regardless of the particular architecture of a system or the skill level of a user.

### Technologies as HI Enablers

As more of the burden for interpreting human communication and for compensating for human error falls on the interface, the performance demands on the underlying hardware and software increase extraordinarily.

Powerful new processors such as Intel's 80486 will be mandatory at the desktop to provide raw computer power. Super-capacity storage devices will be needed to underwrite the tremendous volume of software and data associated with the interface. The volume of software code now being delivered has grown two orders of magnitude in the last ten years. Communication devices and networks will have to accommodate the increased volume of information passing between system and user, and among users. New software approaches will have to be employed to shorten the time required to develop applications and to execute a program, as well as to contain the costs of development. Thus, technology availability and application are central to the success of advanced HI initiatives.

### Candidate Technologies

Object-oriented programming (OOP) is an important emerging software technology.[5] The simple fact is that software development costs must be reduced before anticipated new HI-intensive applications can appear, and OOP is the vehicle to contain costs.

Object-oriented programming departs from action-oriented programming in that it concentrates on describing all the information one needs to know about an object (for instance, a rectangle) in order to accomplish tasks using the object. (In contrast, action-oriented programming takes a step-by-step procedural approach to defining how a task is to be performed.)

OOP has three components: data encapsulation, message passing, and inheritance. *Data encapsulation* is concerned with where one stores the information (attributes) and the instructions for what the object can do (e.g., size, color, placement of the rectangle on the screen). *Message passing* defines the ways objects interact with one another. (For instance, if the rectangle changes size, does the circle care? If one rectangle changes from red to green, should all other rectangles follow suit?) *Inheritance* defines what one object can inherit from another. There can be classes of objects with the same characteristics. (For example, a square is a special kind of rectangle.)

Using the combined power of data encapsulation, message passing, and inheritance, OOP can define complete tasks or actions to be taken by the computer. These actions are executed rapidly, allowing for real-time interaction between machine and user. By "interaction" I mean that the computer is able to respond to the user within a time that closely matches the user's "think" and "act" times. (See reference 1.)

Further, OOP modules are reusable and consistent, which cuts down on the development effort for the next application designed.

Most technologies assume a level of hardware enrichment coupled with software that allows one to tap into the extended capabilities of the hardware. Some of these more important technologies are:

• Voice recognition and response, which will eventually give way to continuous speech recognition and replies. Here, the main constraint at present is an inability (due to inadequate processing power and storage capacity) to handle compression algorithms associated with pattern recognition.

• High-resolution bit-mapped display for presenting excellent quality graphics on the screen.

• Writing-based input (such as handwriting) into the system.

• Scanning for introducing hard-copy characters, text, data, images, and graphics directly into a system. (NCR's scanning scales for weighing, marking, and entering the prices of produce at supermarkets serves as an excellent example of addressing HI needs through scanning.)

• Displays which are smaller in volume and larger in display area, displays which will react to touch, and displays with broad-spectrum and exacting color resolution.

• Multimedia devices for mixing audio, video, high-resolution stills, graphics, and text.

• A data glove and a pressure-outputting (pushing back at you) joystick for direct manipulation of objects on a screen.

• Enhanced and less expensive color printers.

• Gesture recognition for interpreting body language.

• Digital video imaging, which will be able to store 60–70 minutes' worth of viewing material on a single disk.

• High-definition TV, which will find its way into workstation applications.

• Vision tracking devices for initiating action based upon what one is watching.

• Holographic imaging for three-dimensional display and modeling. (Think of a terminal in a furniture store that allows you to "build" a room full of furniture on a screen.)

• Pressure-sensitive pads for writing or drawing directly into the computer.

• Neural nets that will help solve pattern-recognition challenges in artificial-intelligence applications.

• Signal-processing chips and faster processors.

• Huge-capacity magnetic and optical storage devices for handling the burden of image processing and other HI implications.

Additionally, the Integrated Services Digital Network (ISDN) will be needed to carry multimedia, multimodal information to and from workers.

Two other software technology sets will be important to HI. Artificial intelligence will be used to determine characteristics of the user. This information, in turn, will be used by the HI to prompt the user and to tune the system's method of responding to the user's preferences. For instance, would an individual rather hear a communique from the system or see one?

Hypermedia and hypertext will help users navigate through relationships among the various forms of information they handle.

Many of these pacing technologies are imminent. By the end of the 1980s we saw the advent of high-resolution video, NCR's color ATMs, widespread acceptance of compact disks, the NEXT machine, Wang's Freestyle, handwriting-recognition devices, information kiosks, and public facsimile machines. The trick will be to harness the power of appropriate technologies early on in ways that promote meaningful HI capabilities.

### The Phasing In of Technologies

HI will have a special place in the annals of technology, for it is the first application of technology that cares what the user thinks. Based upon an improved understanding of human factors in the psychology of user perception, a number of apparently disparate technologies are being brought together at the workstation to significantly upgrade the human interface by stimulating the human senses. We can categorize these technologies into four functional areas:

output technologies

input or entry technologies

image-management technologies

command technologies.

Some technologies, such as displays and voice, will be used for both input and output. This will cut down on user disorientation and will take us closer to the way we interact with one another.

Additionally, multiple modes of communicating will be supported. For instance, the user could begin inputting to the system from a keyboard, continue by dictating, and finish the same exercise by manipulating a mouse. The system could also reach out to the user with intermixed or simultaneous images and sound.

In the area of command technology, the user eventually will be able to tell the system what to do by speaking a command and/or by directing the system through eye movement or gesturing. Other technologies, such as pointing devices (the mouse), will be used in the interim.

Presentation technology is taking on heightened importance. The display has become the "image of the computer system." The quality of the picture and the speed of graphics management have an impact on the user's overall satisfaction. Many factors are related to the overall effectiveness of a presentation subsystem, including display resolution, convergence, chromaticity, brightness, pixel size, and bandwidth.

The display controller requires efficient frame buffer architecture, digital-to-analog conversion, and memory management, while the software and user interface design provides windowing, graphics, fonts, and overall ergonomics.

By way of minutiae, 640 × 480 pixel resolution for CRT displays and 300 dots per inch (dpi) for laser printers are 1990's de facto industry standards. They will rapidly move to 800 × 600 pixel resolution for CRT displays and 400 dpi for printers.

Much of the aforementioned has already evolved. Large displays for group viewing are available, as are color CRT technology in the range of 1K × 1K pixels. Image video and animation should have just been introduced. By the mid-1990s, color flat-panel displays will provide adequate reflectiveness and brightness and will become an economically viable alternative to CRTs for many applications. Moreover, object-oriented software will allow us to apply parallel system software to the human interface, thus rendering "interactive" entire suites of computation currently thought of as "batch" only.

Before the middle of the 1990s, the HI technology family may include expert systems, voice input, high-fidelity sound, and the abandonment of the CRT in favor of the flat panel. By 1998, 2K × 2K-pixel bit-map displays should be in use, providing as high a quality of imagining as the eye and the brain can register.

With all these advances in process, what might we expect to see as new HI-based applications during the early 1990s? Here are just a few to tickle the imagination.

• The voice typewriter, by means of which we will speak to a computer and our input will be automatically word-processed and a rough draft printed out.

• The ticketing terminal, which will display the view from a selected seat in a theater or a ballpark.

• Real-time check-writing verification based on both the design of the signature and the time it takes the writer to execute it.

### Fulfilling the Promise of HI

What about the technical artisans who must masterfully fashion the HI model of the future and then develop and improve it with new technologies? What are the major challenges they face? What might slow the pace of HI advances?

From a practical standpoint, HI developers and researchers suffer terribly from fragmentation because the subject is so broad, complex, and volatile. There are bits and pieces of development going on in many computer vendors' camps and at universities, but there is no coordinated effort to advance the state of the art. Perhaps some understanding of HI guidelines and a shared base of expertise will evolve as HI products leave the labs and generate excitement among users.

Right now, the HI developers are speculating with regard to what the users want and need. There are no tightly specified requirements for HI approaches, and the developers must acquire an understanding of the users and of their intended (application) environments. Developers must expand their knowledge of how humans think, act, and are motivated. This means having a spectrum of professionals who have never worked together before (e.g., cognitive scientists, graphic artists, and electronics engineers) feeding into the design effort.

The fast pace of HI work precludes efforts to standardize. Prototyping and testing with an eye to dynamic change, rather than designing to specification, seem a plausible commercialization route for HI.

The HI developers need to communicate heavily with application developers, so that their efforts are synergistic and the resulting pieces of a computer system can play together. In general, the more complicated the application, the better the HI must be in order for humans to use the application productively.

Perhaps the biggest challenge of all is to manage HI projects of an intangible nature in a development community that is quantitatively oriented. These challenges may seem overwhelming. Yet we know that the human interface is the key to unleashing the power of computers—the key to raising personal productivity. The developers know this too, and they are pushing forward into the uncharted waters of the future.

The developers will be proclaimed successful when we, the users, are able to do our jobs without knowing or even thinking about the computers that are helping us—an imperative for coping with the information intensity of the next millennium.

### Acknowledgments

I wish to acknowledge the contributions of the NCR Human Interface Center staff, who supplied numerous facts and ideas for this chapter. I also recognize the substantive material provided by Tom Mays.

Special thanks go to Charnell Havens, our professional colleague and the technical writer for this chapter. Her insight and expertise have turned a complex subject into verse that is lively, informative, and a pleasure to read.

### References

1. Thomas Mays, The '486 Shockwave: What the Future Holds. Presented at PC Expo Seminar, June 21, 1989.

2. Charlene Renee Benson, The Use of Single-Page, Direct Manipulation Interfaces in Real Time Supervisory Control Systems. Report No. 89-3, Georgia Institute of Technology, March 1989.

3. William Newman and Robert Sproul, *Principles of Interactive Computer Graphics,* second edition. McGraw-Hill, 1979.

4. S. K. Card, T. T. Moran, and A. Newell, *The Psychology of Human-Computer Interaction.* Lawrence Erlbaum Associates, 1983.

5. Bruce Shriver and Peter Wegner, eds. *Research Directions in Object-Oriented Programming.* MIT Press, 1987.

# 12

## *In a Very Short Time: What Is Coming Next in Telecommunications*

*Robert W. Lucky*

When I visit my parent's home—the house in which I spent my youth—I often pause to gaze reflectively at the old, black, rotary-dial telephone in the hallway. It is like my picture of Dorian Grey. That ancient phone is the very same instrument I knew as a teenager so many years ago. Whereas I feel ever youthful, it has aged markedly, reminding me of both how little and how much telecommunications has changed in my lifetime.

Perhaps that old phone is like a well-kept antique car that functions just as it did when it was new. From the viewpoint of the owner of such an antique, its modern equivalents offer few, if any, necessary or attractive new functions. But one has only to drive the car onto the highway or rotate the dial of the telephone to realize the changes that time has brought. The infrastructures of communications and transportation have, at enormous expense, been revolutionized. Vehicular traffic whizzes by at high speeds on superhighways that fly through cities on elevated platforms and interconnect in dazzling spirals. It is the same with the telephone network, but in either case the function of the network as seen by the user appears the same as it always has. The telephone and the car seem unchanged in what they do for us.

Behind the deceptive facade of the black telephone, out there in the network, a revolution has been growing. Like the accumulation of spring rains in the distant mountains, the effects have so far only trickled to the user community, while the potential for dramatic revision in the way we see and use communications has been building relentlessly. Indeed, changes are all about us now—facsimile, data communications, cable television—yet they have arrived so modestly and in so piecemeal a fashion that we view them with complacency. Through most of this century we became accustomed to the quiet infiltration of change, but in the mid-1980s the pace accelerated. The world of communications is churning with new ideas and new directions.

## The Historical Evolution of Telecommunications

A century ago the United States was spanned by an all-digital telecommunications network designed for the purposes of telegraphy. At that time the largest corporation on earth was the Western Union Company. Communication was effected by expert operators, tapping with a mindless virtuosity at telegraph keys. The wires hummed with the dots and dashes of the new language. Users couched their messages in "telegraphese," conserving letters, and carried them down to the nearest telegraph office, where they might wait patiently for a reply from the mysterious black wire disappearing like a railroad track into the horizon.

As the telegraph system grew and thickened its branches and twigs throughout the nation, inventors raced frantically to perfect an instrument for voice transmission. The famous, the infamous, and the obscure pretenders all had telephones, but when the confusion finally cleared a Supreme Court decision awarded the patent to Alexander Graham Bell. The crux of Bell's invention of the telephone in 1875 was the use of analog transmission—the voltage impressed on the line was proportional to the sound pressure at the microphone. It seemed that digital transmission was not so desirable after all. Although some people felt that written transcriptions made telegraphy far more desirable than fleeting voice conversations, Bell's telephone eventually swept the land. The early digital network—the telegraph—gradually faded from view.

Now that voice communication has almost totally supplanted the written telegraph format, we might consider the relative advantages and disadvantages of these media. Perhaps the most important factor in voice transmission was the elimination of the intervening translator. For the first time an untrained person could deal directly with the communications channel. In addition to the all-important matter of convenience, the transmission took place instantaneously, and permitted a dialogue between the participants. The voice was less formal, and it encouraged elaboration (which had been, by convention, eliminated from telegraphy). The advantages of the telegraph medium were much more subtle, and it took many years for engineers to come to appreciate the possible robustness and speed of the digital format. Meanwhile, there was to be a century of analog transmission.

The goal in the early growth of the telephone network was to connect the nation. In fact it took almost half a century to accomplish this transcontinental connectivity, but in the mid-1920s it became possible to place a call between any two telephones in the country. Still, however, the telephone was a luxury that was found in a minority of American homes. The driving force in telecommunications then became the achieving of universal telephone service—everyone should have a telephone. The telephone had turned from a status symbol into a social necessity, and the establishment and regulation of the telephone monopoly were intended to promote that social good. The telephone was seen principally as a lifeline, a means to chat with neighbors and to summon emergency help. The idea of that telecommunications line as a link to a world of information still lay in the future.

When the goal of universal telephone service became virtually a reality, in the 1970s, some of the wind went out of the sails of the telephone pioneers. Some said the void after the fulfillment of that inspirational goal was akin to a loss of religion. Now the question was: What next? It was no longer obvious, but the concept of telephony was undergoing redefinition. Computers were beginning their spectacular growth, and the first tremors of their needs for data communications were being felt. Voice telephony had become easy, and long-distance traffic was burgeoning.

About this same time there was a shift in the mood of the government and the public about how telecommunications should be administered in the United States. Monopoly had served the country well during the spread of telephony, but now there were strong social and political forces tending to favor increased competition. The Carterphone decision in 1968 opened up the network to the connection of so-called "foreign" (non-Bell) terminal equipment. A few years later the Federal Communications Commission approved the petition of MCI to compete with Bell in providing long-distance transmission. Before long the Justice Department instituted a suit against the Bell System under the anti-trust laws, which eventually resulted, in 1984, in the dismembering of the Bell System into separate, competing parts. This divestiture was administered under what was called the modified final judgment, which even today remains a living agreement under constant dispute. Other nations took note of the introduction of increased competition in the United States, and both England and Japan began privatizing their telecommunications carriers.

Over these last one hundred years of telephony a number of technological innovations provided the impetus behind universal telephone service. The earliest transmission method was over open wires. At first the wires were strung between poles, but the darkening of the skies from the plethora of wires in the large cities led to their eventually being housed in buried conduits. These open wire systems transmitted a dozen simultaneous conversations through the use of frequency-division multiplexing of the 4-kilohertz analog voiceband channels. In this early period of telephony, which lasted nearly fifty years, switching was accomplished, largely manually, by an exponentially growing number of telephone operators.

Around the end of the Second World War there were a number of extraordinary inventions pertinent to telecommunications. The great wartime refinement of radar led to microwave radio transmission, and in the late 1950s telephone traffic began to be carried through beams of radio transmission, marching from tower to tower across the country and carrying thousands of conversations. The invention of the transistor and the conception of the digital computer changed our world in dramatic ways, but in telephony their largest impact was felt in the introduction of stored-program electronic switching. The fear that the number of telephone operators required for the growing phone system would exceed the population proved groundless. Machines could do the job.

That burst of creativity in the second half of the 1940s also included the brilliant insight into the nature of information achieved by Claude Shannon. His paper "A Mathematical Theory of Communication," published in 1948, created a philosophical revolution in the way information and communication were understood. Shannon raised the thinking out of the physical muck of dots and dashes into the virtual world of information. He answered such questions as: What is information? How much information can be carried by a communications channel, such as a wire? In answering these questions Shannon arrived at the startling conclusion that communication channels could convey information at a rate that he called "capacity" with a vanishingly small probability of making an error. In later years his work led to the error-correcting codes that are used everywhere in communications today.

In the late 1940s, there was a slow realization of the advantages of pulse-code modulation (invented by Reeves in 1939) as an alternative method of sending a voice telephone signal not as an analog signal, but as an encoded sequence of pulses not unlike telegraphy. Thus began the most important modern trend in telecommunications: that of digitization. The first digital carrier system, the T-1 carrier, was installed by the Bell System beginning in 1961. In this system a voice channel was encoded into a stream of 64,000 bits per second. Subsequently, 24 such channels would be time-interleaved, eight bits at a time, onto a single physical wire. The beauty of that approach was only gradually appreciated by engineers. It was that noise and distortion were not allowed to accumulate, since the ones and zeros could be regularly restored (i.e., regenerated) by a succession of repeater sites along the transmission line. The wheel had turned full circle; telegraphy had something to offer after all.

The initial digitization of the telephone plant was confined to interoffice, metropolitan transmission, but gradually the benefits of digitization began to spread. The first digital switching offices, which used time-division multiplex methods to route the ones and zeros of different channels in their respective directions, were installed in the long-haul plant. In retrospect this 4A Electronic Switching System was a daring engineering accomplishment, but one that proved both economical and capable. In the 1970s digital switching spread to the local offices. By the early 1980s much of the telephone plant was digital, but the long-haul carrier systems—the microwave systems—were still largely analog, as was the copper wire that provided the last few thousand feet of transmission into the home.

The digitization of the telephone system was greatly aided by its confluence with another technological evolution, that of lightwave transmission. In the beginning of the 1960s the head-lines in the communications world belonged to satellites. Following an idea articulated in fiction by Arthur C. Clarke, and pursued technologically by John R. Pierce and his associates at Bell Laboratories, relay by satellite was demonstrated by passive reflection off an orbiting balloon known as Echo. Shortly thereafter, the Telstar satellite was launched to provide active relay, and for the first time the nation witnessed live television relayed from Europe. The traditional communications horizon at television frequencies—line of sight—had been extended across the continents.

While satellite technology was absorbing the attention of communications designers, the invention of the laser gained little notice. Satellites were brought almost immediately to communications applications, whereas the laser waited two decades before it became a dominant force in telecommunications. But in 1971 the first optical fibers with a transparency suitable for communications use were made. These hair-thin threads of glass had at that time an attenuation of about 20 decibels per kilometer. Roughly speaking, that meant that it was possible to "see through" about one kilometer of fiber. This feat caused an immediate increase in efforts to send communications signals by light wave. In 1981 the Bell System installed the light-wave "Northeast Corridor" from Boston to New York, Philadelphia, and Washington. Although this was the first large-scale application of long-haul optical transmission, an earlier experiment for intracity transmission in Atlanta had proved that optical fibers could be a commercially viable means of communication. It was a brave experiment indeed; who would trust their communications to little pulses of light, after a century's refinement of electronic means?

The light-wave fiber transmission systems were inherently digital. It was too difficult to transmit with high fidelity many analog signals over a single fiber without mutual interference. Only the robust ones and zeros, represented by light and no light, seemed capable of providing reliable communications. As light-wave systems were gradually introduced into the long-haul telephone plant to replace microwave radio systems, one of the most important benefits was the digitization of the long-haul portion of the network. By 1989 the long-haul network of AT&T was essentially digital. At long last, ones and zeros were the natural language of the network. If people insisted in talking over the network, then their voice waveforms would have to be changed to digital streams. Indeed, all information henceforth would be carried in a digital format. The natural language of computers, bits and bytes, had prevailed.

The digitization of the telephone plant coincided with an increased emphasis on data transmission. This was indeed the beginning of the information age, and the traffic on the network began to reflect this person-to-computer and computer-to-computer communication. In the early 1960s the data traffic was handled exclusively by modems that converted the data to a voice-like analog signal. As the network itself became more and more digital it was easy to predict that modems would no longer be necessary, but their sales increased more rapidly than ever. Today millions of modems still carry most of the data traffic, since the local-loop portion of the network remains undigitized. However, we are poised on the brink of end-to-end digital connections, and perhaps before long the necessity for modems will be seen as a passing, awkward phase of telecommunications.

### Telecommunications Trends Today

Relative to that black telephone in my parents' home, the services that technology has brought may not seem so significant. Perhaps this is because most of the technology went toward universal service and toward decreasing the cost of communications. Both were very important social achievements. Today everyone has a telephone, and for many users the cost of a call no longer inhibits communication. When I went to college, many years ago, I never called home. It was a period when the very words "long-distance call" evoked ominous worries of impending tragedy. Today my children call from college nearly every day. It is accepted, ordinary behavior. Thus, the quantitative change of universal telephone service became so large that it became a qualitative change in our behavior.

Cost as a barrier has been eliminated for many purposes, and distance has assumed a different psychological scale. Satellites have made international calling from anywhere to anywhere commonplace. In London I put a few small coins in a public phone and talk to my New Jersey office or home for a minute or so. No big deal. As I write these words, the optical fiber systems spanning the Atlantic and the Pacific creep to within miles of their destinations across the seas. So far as communications connectivity is concerned, the global village is already here.

*Accessibility*

Access to this international communications infrastructure is much, much easier than in the past. We are seldom out of sight of a telephone. Chances are very good that if you look up from this book you will be able to see a telephone. With phones everywhere, and with the convenience of credit card calls, we are always in touch.

Accessibility has also been promoted by portable and mobile telephony. We have been coming at portable communications both from the large and from the small. The large approach is cellular radio, with an investment in network infrastructure supporting radio cells that shrink in size as necessitated by traffic growth. As the cells get ever smaller, we approach ubiquitous personal communication. The global infrastructure of mobility is also enhanced by paging systems and by such mobile systems as the "airphones" in commercial airplanes. Meanwhile, a more limited form of portability has been evolving from the small, as is evidenced by the growing number and power of cordless telephones. Today it is possible to call from your back yard using your home-based cordless phone. Why not also from your local grocery store?

The accessibility of communications has to do with more than just the ubiquity and mobility of telephones, since communication involves both people and instruments. It is still true today that we make calls to *places,* rather than to people. With people in constant motion, this means that calls and people do not intersect as much as we would like. Thus, one of the significant trends in recent years has been the rise of telephone answering machines. More than a decade ago the Bell System developed a voice storage system which provided centrally the equivalent of an automated answering and messaging service, but it was forbidden to market the system because of a government ruling. Instead, the same service grew from the periphery of the network on a more personal level. In spite of the jokes and the frustrations they inspire, answering machines offer a valuable social service: the time-shifting of communications in at least a limited format.

Another form of messaging that has blossomed in popularity today is facsimile transmission. Invented decades earlier, facsimile was never popular until the mid-1980s. Many inventions occur before their time, so to speak, and this was one of them. Today's popularity of facsimile demonstrates the influence of international standardization, as well as the market appeal of being able to manufacture a product that meets a critical low threshold price. A futurologist of only a few years ago might have ignored the "fax" completely, but today it is so significant that the majority of the telecommunications traffic between Japan and the United States is in this form. The preponderance of facsimile traffic in the trans-Pacific corridor is undoubtedly stimulated by both the large time shift between the two countries and the Japanese fondness for graphical expression. However, even in the text-oriented United States the fax machine is now becoming a standard office appliance, and we wonder how we lived so long without it.

The economics of telephone communications today are dominated by the cost of providing local access. The pair of copper wires between the home and the local distribution point, the subscriber-loop multiplex system (which may combine the voice channels of several dozen homes), and the portions of the central-office switch and long-distance access which are attributable to the individual home are assets whose cost and maintenance must be amortized through the telephone access charge. It has long been a truism that communications is a "last mile" problem. Technology offers greater and greater economies of scale in the network itself, but few in the local loop. You are on your own in getting to the network. No technological means of making this access cheap are in sight.

### *Transmission*
Another important dimension of communications is the capacity of the interconnecting links. This is usually expressed as bandwidth in analog systems, or as data rate in digital systems. In either case we can think of capacity as the relative freedom of expression. For a century the capacity of telephone links has been set as the voice channel as it was defined long ago—about 3 kHz. The speech quality that has traditionally been rendered by the telephone is thoroughly intelligible for a broad range of speakers, though it is rather far from high fidelity. When modems are used to transmit data over these voice channels, data rates in the range of 2400–9600 bits per second are obtained with increasing sophistication and cost of the modem. Rates as

high as 19,200 bits per second can be achieved with very sophisticated modems, and this rate seems very near to the theoretical limit as expressed in Shannon's channel-capacity equation.

Much more capacity is required to convey television. A network-quality television channel needs a bandwidth of about 2 megahertz (or, digitized, a data rate of about 45 megabits per second). It seems that a picture is indeed worth a thousand words! In any event, it takes about a thousand times the communications capacity of a voice channel. The existing telephone connections cannot carry this capacity to the end user in the home. Thus, one of the most important developments of the 1980s was the gradual rewiring of the country with community-antenna-television (CATV) cable. A high percentage of homes in the United States are now wired for CATV, or are in close proximity to a CATV distribution cable. This separate infrastructure for television distribution is one of the significant factors shaping the future of telecommunications in the United States.

The CATV distribution system supports about 100 television channels. However, it is strictly a broadcast system. That is, everybody gets the same signal; there is no individualized switching. Thus, most homes are wired with two incoming cables: one, for the telephone, is a narrow pipe which is individually switched to a densely interconnected network; the other, for CATV, is a wide pipe connected to an immutable, common stream. The designers of each of these pipes obviously have dreams of overcoming their respective disadvantages.

The irresistible force in capacity today is transmission by optical fiber. The existing media of microwave, satellite, and radio transmission are essentially used up. The microwave links entering large cities converge in a traffic jam. The satellite parking places in synchronous orbit are filled, and the spectrum allocation is crowded. The radio spectrum for mobile communications is a precious resource that requires very clever management and reuse, such as in the cellular system. But the capacity of optical fiber seems limitless.

Since optical fiber first became feasible, in the early 1970s, its capability to carry communications traffic has doubled every year. To be precise, the product of the bit rate obtainable experimentally on an optical fiber and the distance the data is transmitted in the experiment has followed this exponential, compound 100% interest curve for twenty years. How far we go on fiber, times how fast we go, doubles each year. Today the light-wave record is around 16 gigabits (billion bits) per second over a distance of about 80 kilometers. (The factor of distance transmitted has only to do with the necessary spacing of repeater sites in manholes and in undersea locations.) As a point of comparison, this rate of progression is twice the rate at which semiconductor electronics have shrunk in size and increased in circuit density.

Today the commercial fiber systems that span the United States transmit at a rate of 1.7 gigabits per second per fiber. This is equivalent to about 25,000 voice channels on each slender fiber. The country is already wired across its backbone with these fibers, and each year we learn better and better how to use them. Fortunately, the progress in attainable capacity has been achieved not by making better fibers, which would have required a rewiring of the country for each improvement, but by making better laser transmitters and photodetector receivers. By changing the electronics in the repeaters, we make the capacity larger and larger.

Silica fibers today have an attenuation of only about 0.25 decibel per kilometer, which is almost as transparent as is theoretically possible in this material. Roughly speaking, again, this means that we could "see through" about 60 miles of today's fiber! Although more exotic materials are known which have a theoretically lower attenuation, progress in making actual working fibers in these materials has been very slow. It might be possible to make an optical fiber that could convey signals across the Atlantic Ocean without needing a repeater, but it is not obvious that this would be of great benefit to the economics of terrestrial transmission.

So the fibers already buried in the ground are an investment in the future. How much traffic could they carry? How much longer can we continue to double their capacity each year through innovations in transmitters and receivers? The answer is "a long time"; the end is nowhere in sight. It seems that the fiber's intrinsic information-carrying capacity is almost a factor of 1000 away from where we are now. Instead of handling 25,000 voice calls simultaneously, that same fiber might in the future handle the equivalent of 25 million calls. Or it might handle 25,000 simultaneous television signals. So far we have only transmitted light of one particular wavelength over the fibers. We have used only about 0.1% of the available spectral window of transparency. It is as if we had developed AM radio but allowed only one station to transmit. When we ask for communications bandwidth, it is already buried in the ground asking to be used. The future lies there waiting to be discovered. The rate of exposing this potential through increased research knowledge is, as was noted above, 100% compound growth per year. But the annual rate of growth of the communications industry is only about 8%. This industry growth is proportional to the growth in traffic. How can we reconcile this dramatic difference? Several scenarios represent possibilities for the future. Will long-distance transmission become free? That seems unlikely, as communications is in the final analysis a service industry dominated by "people costs." Another alternative is that the traditional voice-band channel will widen without becoming appreciably more expensive. The buried fibers might stimulate the need for wider-band communications services, such as image and video transmission.

Even though the fiber transmission systems promise greater bandwidth, they have not yet delivered on this promise. The fibers today are being installed for other reasons. They offer a quick injection of capacity to the new long-distance carriers that are just building networks at this propitious time. For the established common carriers the fibers offer a bulk upgrade to a digital network by replacing the older analog systems. The fibers also appear to deliver a higher quality of service with less noise (really because they are digital) and higher availability. Finally, the fibers are being used to help solve a problem they themselves created: the vulnerability of the network to catastrophic outages. Construction plows occasionally cut buried transmission lines, but in the past only a small fraction of the system's capacity would be affected. Today, with so many eggs in one basket (the fiber), such an outage is truly catastrophic to the network. Thus, more fiber routes are needed just to offer physical redundancy to the routes already installed.

Finally, the fibers are like superhighways spanning the country, but they are highways without efficient on and off ramps. Those large pipes are out in the network, but they do not reach the homes. At home, we are forced to drink from those fire-hoses of bits through the straws of our narrow-band copper connections. One of the most significant questions for the future is whether the fiber will reach the home. If it does, what will we drink from that firehose?

## Switching

For decades connectivity meant whom you could reach with your phone, and it was simply accomplished by an operator pushing a plug into a board to connect the wires physically. Stored-program electronic switching replaced the plug and the operator with computer-controlled relays. By this time everybody could talk to everybody else. Now people began to demand more customized services, such as call forwarding, speed dialing, and call waiting. To an amateur programmer such features sound easy to provide, merely by writing a few lines of software. The reality is quite different.

Today's central-office switches, like AT&T's ESS No. 5, are among the most complicated systems made by mankind. The switch hardware is entirely digital, and instead of connecting people by putting wires together it routes bytes of data using time-slot interchange. For example, the byte of bits representing the latest fragment of your speech waveform is moved in time to a different position in the stream of multiplexed data, at which time your listener will be electronically watching for his next byte.

The real complexity of the current digital switches lies in their software. The ESS No. 5 software comprises about 2 million lines of high-level code. Of the portion that actually runs in the switching system (some of the code is for support of the programming environment), perhaps two-thirds is for maintenance—what happens if such-and-such should go wrong, that kind of thing. In order to develop this much code, thousands of man-years of effort, costing about a billion dollars, must be invested. It is a huge investment, and one that is not easily ignored in planning for subsequent generations of central switches.

The digital switches are a triumph of modern systems engineering, but they bring their own limitations. One is the difficulty of adding new features. In concept a new feature is created by adding more software, but when the new software has to interact with a million lines of existing code the complexity is overwhelming. Changing the code is often described as being like pulling a strand from a plate of spaghetti. Thus, new features come about much more slowly than users might wish, and there is always a substantial backlog between what the market demands and what the software developers can produce. Switching today is very much the business of managing complexity. What a far cry from the public image, not so long ago, of the telephone worker as a spike-booted lineman!

The other problem with the digital offices lies in their limited capacity. With fibers of increasing data rates being installed in the transmission links, the offices could become more and more overwhelmed with high-speed bit streams. The hardware architecture of the office demands that the switch look at individual bits in the incoming streams in order to do time-slot interchange. Separating droplets in those firehoses is becoming increasingly difficult, especially since the rate at which electronic circuits are being increased in speed is only half the rate at which optical speeds are being increased.

Engineers are pursuing two approaches to a next-generation switch: fast electronic *packet switching* and *photonic switching*. Of these, packet switching is the one closer at hand. Whereas the current digital switches are circuit-switched, in packet switching each package of data carries its own address, and can flow individually through the network much as a letter flows through the postal system. At any routing point an electronic circuit can read the address on the packet and decide which way the packet should travel, just as a sorting clerk in the post office might check the zip code before throwing a letter into one bin or another. Experimental switches that work on this sorting principle are being constructed to handle data at rates orders of magnitude higher than the current voice-band rates of 64,000 bits per second.

Some engineers envision a network which is entirely packet-switched. Voice, data, and images would look alike to the network, all being represented by successive packets of data. In many ways this is an appealing scenario, but a closer look reveals difficulties. One difficulty is that voice samples must arrive at regularly spaced intervals; otherwise the voice sounds "chopped up." Since successive packets can take different paths through a network, their arrival times (and even their ordering) cannot be guaranteed. Computer data does not have this same sensitivity. Another difficulty is that as we look deeper into the network hierarchy, where more and more packets are converging at central switching points, the traffic becomes very heavy and the switch must work faster and faster—perhaps faster than electronics is capable of working. (Imagine, in an analogy, the traffic at the airports which serve as hubs for packages of people in airplanes—Atlanta, Chicago, Denver, and so on.) For these and other reasons, it is likely that switching in the foreseeable future will evolve as a combination of circuit switching (as in current telephony) and packet switching.

The other approach to switching is to give up on electronics and turn to optical means of steering the streams of fast-flowing bits which converge on switching centers. At present there is considerable interest in photonic switching as the long-term solution to the bit-rate crunch at switching centers. Physicists are experimenting with arrays of new devices that are like optical transistors—they can be turned on or off (transparent or opaque to a main beam of light) in response to either an electrical or an optical control signal. We might imagine information entering the switch in many parallel beams of light, each of which carries a high-speed data stream. The parallel beams of light would pass through planar arrays of these photonic on-off switches, eventually emerging rearranged. Nowhere is the light-wave signal converted to an electronic signal. In spite of the considerable research interest in this approach, it will probably be several years or more before prototype systems are constructed.

### Networking

The physical connecting of wires, or even of bits, is only part of the important issue of connectivity in today's communications. The information age has brought the larger issue of logical connectivity among computers and their users. In contrast, voice conversations between people are easily networked.

Through centuries of experience, we have developed conventions for information exchange between people. You ring my phone, I say "Hello," you say "This is . . . ," and so on. We agree to speak the same language, though the phone itself will handle any language transparently. We can speak as fast as we like, but if you do not understand something I have said, you ask me to repeat. We do not even consciously think about these protocols that guide human dialogue.

Computer communication also has protocols. Moreover, computers are very fussy about what they understand; they treat everything entirely too literally. Nor can they speak as fast as they would like, since communications lines are a bottleneck to data. One would think that we would have agreed long ago on protocols for dialogue with and between computers. Unfortunately, we have not exactly done so. Though virtually everyone I know in the engineering and science professions has a computer or a computer terminal on his or her desk, the fraction of people to whom I can send computer mail is very small. Our computers do not speak the same language. Possibly our computers are talking to non-connected networks. There is not even a telephone book so I can find who is out there! The situation resembles voice telephony at the turn of the last century, when each city had its own telephone company. Surely we must do better in the future.

Why is it so difficult to agree on standards and protocols for computer communications? It is partly a question of the mismatch between the pace of technology and the time that it takes large groups of humans with conflicting motives to reach agreements. Technology simply has been unable to wait for the agreements to be made. Therefore, networks have been built before the detailed standards have been available. Although this meant that incompatibilities would develop later with other networks, it did allow many flavors of networking to develop, and the systems themselves were at least welcome in the marketplace.

When a computer network is localized, all the interfaces can be under the control of one manufacturer, who can see that all connected users play by the same rules. Thus, local-area networks (LANs) are a world in themselves, and a world that has proved to be enormously useful in the present stage of computer evolution. Ethernet was the first widely accepted LAN, transmitting data at 10 million bits per second on a coaxial cable of up to several thousand feet in length. User data is packetized, and a protocol enables a user to capture empty time slots for his packets. There is no central intelligence which manages access to the loop, merely an agreement about how the passive cable loop is to be used by all. In fact, Ethernet is very much like a human conversation around a large table. If you want to speak, you wait until there is a lull in the ongoing conversation; then you speak. If interrupted by someone else deciding to speak at the same time, you retreat and try again later.

In recent years LANs have sprung up throughout business, academia, and government. Powerful workstations sit on desks, promising instant computer power to their owners, while the files that they manipulate reside elsewhere on the LAN, where they are managed and maintained centrally. It is, for the moment, the best of both worlds—distributed processing, but centralized file servers. Moreover, the speed of data within the LAN is orders of magnitude faster than users had to endure when talking through modems and the telephone network. Users feel spoiled with this easy, fast access to communications. Too bad it only reaches within the building or campus. The whole world should be this way.

In a LAN, the medium itself (e.g., the coaxial cable) is passive, and all of the intelligence is distributed to the periphery. In contrast, in the outer world of wide-area networking there is much intelligence within the network. A decade or so ago it appeared that intelligence was an expensive commodity that belonged inside the network, where it could be shared by many users. Thus, computer networks were designed to use central processing and storage within the network. The personal computer changed all that, and the ensuing revolution caused some of the intelligence to migrate from inside the network to the periphery. Today there is much discussion about where intelligence belongs, and about what functions intelligence should serve in a communications network. As an important example, consider the need for translation protocols between differing computer networks—a situation analogous to requiring a human operator to translate from a foreign language to English. Where should the translation points be, and who should own them?

There is another reason why it is hard to agree on standards for data communications: computer networking is intellectually much more complex than voice telephony. Today there is no universal consensus about a suitable abstraction for the channel. In what terms should we think of the communications medium? For example, we might think of it historically as a wire, with all the physical properties of a wire. The wire is like a pipe for bits. The bits come out in the same sequence in which they were put in, but they are delayed by some fixed amount. However, what happens when there is an error? Furthermore, computer software systems do not like to have to deal with individual bits. For these reasons we might prefer to deal with the concept of a file, as we do in our dealings with computer operating systems. Let the channel worry about how the file is assembled, transmitted, and checked. For some communications purposes, this would be appropriate, for others not.

In the last decade or so we have seen a gradual shift in the way networks are engineered. In addition to the rising complexity imposed by computer communications, there are important social and political changes. When the Bell System had a monopoly in the United States, it could dictate standards through its own designs. That is no longer the case; now these standards must be evolved through international standards bodies for telecommunications. The role of these standards bodies has become preeminent. Progress in determining the future of networks is no longer a matter of inventions by individuals or even of teams. It is more a matter of the integration and socialization of evolving knowledge through layers of committees that decide the future, much as the legislature eventually determines the will of the people and enacts suitable policies through the passage of bills.

### The Near Future—ISDN

It is of course impossible to predict the future of telecommunications with any degree of certainty. Twenty years ago it would have been impossible to foresee the rise of optical communications and the blossoming of the personal computer. On the other hand, there is a window on the future—perhaps about a five-year period—that is already being shaped in the research and development laboratories and in the standards bodies. In this semi-predictable, near future, one of the most important themes is the Integrated Services Digital Network (ISDN). ISDN is particularly important in that it will provide a framework for the evolution of new telecommunications services.

ISDN is an end-to-end digital service which is standardized throughout the world. We might think of ISDN as a digital socket on the wall, like the electric power socket, that transmits and accepts digital streams of bits—a kind of universal service for data. To be more precise, the basic rate interface in ISDN accepts two separate 64,000-bps circuit-switched streams (called B channels) and one 16,000-bps packet-switched stream (called a D channel). Thus, the basic-rate interface is usually described as 2B + D. There is also a primary-rate interface at a higher rate that is equal to 23B + D.

What is new about ISDN is its end-to-end digital capability, its much higher data rate (modems achieve only about 4800 bps) and the flexible way in which it can be used. In particular, the user has access to a packet-switched channel which he can use to send messages to the network itself. These services have been in trial since about 1986, and ISDN itself first became public in the United States in 1988 with the shipment of several hundred thousand lines of access. Over the next five years ISDN will undoubtedly grow in the form of islands around the central offices in which it is installed and the private networks in which it is featured. After that period, the islands should start to melt together, and the private exchanges and LANs should have ISDN gateway capability, making the service universal.

A few immediate benefits of ISDN will be apparent to early users. Telemarketing services can make use of the calling-number information that is fed to the receiving terminal during call setup. In other words, the receiving computer system knows who is calling before the phone even rings. All of the pertinent file information about the account of the caller may appear on the screen as soon as the call is answered. During the further course of the transaction, the account clerk can communicate data with his local computer, or even with the computer of the caller, on the same ISDN line, and without a modem. ISDN combines voice telephony and data transmission on the same channel. No longer do the computer terminal on the desk and the phone on the same desk have to live in different worlds.

### Audio

In ISDN each of the B channels is a 64,000-bps digital stream. This is the data rate at which voice channels are encoded for so-called "toll-quality" transmission in the heart of the network today. However, remember that today these same voice channels must reach the heart of the network in analog format over a 3-kilohertz channel, whereas in ISDN they are encoded immediately at the originating telephone itself, and thus never encounter the limitations implicit in the narrow-band, analog connection. The lifting of these limitations makes possible several new audio services which may be offered in the early years of ISDN development.

The first of these is high-fidelity voice. The 3-kHz standard for voice bandwidth has been with us for a century, and we are quite used to it. Voice-quality tests through the years have shown that 3 kHz is the minimum bandwidth at which speech retains its intelligibility for a wide spectrum of the population—including especially women and young children, who usually have high voices. In ISDN the 64,000-bps data rate is sufficient to encode voice digitally with modern coding algorithms at a bandwidth of 7 kHz. This is analogous to being able to turn up the treble control on your hi-fi system from a very low setting to a middle-range setting. The voice quality sounds much more natural, and suddenly you may realize how much you have missed for all these years of conventional telephone-quality speech. Conceivably, high-fidelity speech will make it easier to deal with other people, since often we depend not so much upon the logic of the spoken words as upon the tone and the nuances of expression.

Another voice service that ISDN will make possible is speech that is encrypted for privacy. Today there exist secure telephones that are made expressly for government applications. These telephones are quite expensive, and their sound is so poor that it is sometimes hard to recognize the identity of the speaker. The reason today's secure phones are expensive and of poor quality is that the only way to guarantee privacy is first to digitize speech and then to encrypt the resulting digital stream. In today's telephony using analog connections, the maximum data rate obtained by modems is 4800–9600 bps. When voice is digitally encoded at these low data rates, the resulting quality of the speech is poor.

With ISDN's 64,000-bps data streams, the sound quality of speech is excellent. Furthermore, no modem is necessary. Thus, security is relatively easy to provide, and high fidelity is attainable should we choose. As in any secure system, however, the trick is not really in the encryption algorithm but in the key-distribution method. (Safecrackers are sometimes said not to rely on their fingertips but upon finding the combination pasted underneath a drawer in a nearby desk.) Thus, there is much more to a secure telephone than just an ISDN telephone with an encryption chip. Nevertheless, in the world of microcomputers the kind of intelligence required for key distribution is relatively inexpensive to provide in a terminal that necessarily includes other intelligent features.

The market for secure telephones is variously estimated at anywhere from negligible to a billion-dollar industry. Everyone wants privacy, but everyone seems to expect that it is already there. If it is not there, then it should be, and at no cost to the subscriber. In fact, most people seem blissfully unaware of their lack of privacy in ordinary telephone calls, even those made over the cellular radio systems (at least in the United States—this is obviously not true in some other countries). The fact that the intelligence community can intercept the microwave transmission links seems little reason to pay a premium for guaranteed security. However, there is a compensating factor in the new interest in viruses and computer security, which is making system designers conscious of the costs of vulnerability. Market studies have been historically weak in being able to predict accurately what a customer is willing to pay for a given service, so at this time it is difficult to predict the desire for privacy at a price.

Still another service made possible by ISDN is high-fidelity music. The compact disk standard for audio encoding requires more than a million bits per second. Obviously ISDN does not come close to that rate, but recent advances in coding make very good approximations to the compact disk's quality at 128,000 bps, which is the sum of the two B channels available in an ISDN connection. Who would want high-fidelity music transmitted over a telephone line? Perhaps nobody, but I am reminded that the inventors of the optical disk thought that it would be used for video recording. (Who, it was asked, would want to waste all that capability on mere audio?)

### Image Transmission

Images are assuming a much more important role in telecommunications. The higher data rates in ISDN will hasten the spread of image-based applications. For example, facsimile transmission will be aided in quality, speed, and convenience. Something called a fax-phone, or some such name, will combine the functions of a telephone and a fax machine. You will be able to send someone a fax while you are speaking on the telephone. The fax will use one of the B channels of ISDN while your voice uses the other.

The quality and speed of facsimile will be improved through the adoption of a digital facsimile standard called *group 4 fac-*

*simile*. In this standard the scanning of the document can be done at as many as 400 lines per inch—the quality of glossy printed material. Depending on how well the content of the document can be compressed by the coding standard, transmission of documents can take from one to eight seconds.

Document retrieval and still-image transmission are similar to facsimile in the way they are achieved through coding and digital transmission. A typical picture might be digitized in a $512 \times 512$ grid of picture elements, which is 256,000 pels. The algorithms used for picture compression remove the redundancy in the picture and achieve a reasonable quality for color pictures at 0.25 bits per pel. Higher-quality reconstruction of color pictures requires about 2 bits per pel. Thus, the range of times for transmission on an ISDN B channel is from one to eight seconds. For a black-and-white printed page, about a second is needed to bring a readable image onto a screen. Thus, ISDN will not quite achieve the browsing speeds that we would really like to have to leaf through documents, but it comes reasonably close.

The last frontier in image transmission is face-to-face video. Although everyone assumes that the videotelephone represents the future of communications, it was actually marketed as a product by the Bell System in 1971. Few customers were found who were willing to pay the $100 or so a month for the videotelephone, and the service was soon discontinued. In those days video transmission was very expensive, as were video cameras and equipment. Today the videocassette recorder has made video equipment relatively inexpensive, and ISDN will make certain wide-band data rates inexpensive. Moreover, modern coding methodologies for image compression are achieving reasonable-quality video for face-to-face application at the ISDN rate of 64,000 bps. This is "talking head" quality video, not network quality.

One possible way in which the videotelephone service could be implemented on ISDN would be through a plug-in board on a personal computer. In fact, this is rather a generic way in which the other services mentioned here could be packaged. For the videotelephone, the video image of the person could appear as a window on the workstation screen. Again, the beauty of ISDN is the logical combining of a multiplicity of data streams, one of which in this case would carry motion video.

### Multi-Media Conferencing

Conferencing among people at a distance without the necessity for travel has always been an elusive goal for communications. Today such conferencing is accomplished with speakerphones, and to a much lesser extent with special video conferencing facilities. Neither is completely satisfactory in replacing face-to-face meetings. The speakerphone is convenient and relatively ubiquitous, but the medium of audio itself is limited, and conferences among three or more locations are almost impossible to administer. There are no cues as to when a person can talk or as to who is talking from where.

Videoconferencing would seem to be a better substitute for face-to-face meetings. However, it is quite expensive and can only be done from special locations that are permanently equipped with the necessary video equipment. These facilities must typically be booked far in advance, so there is little convenience in the present service. The meetings themselves have a different flavor than face-to-face meetings; they are more formal, there is little opportunity for side conversations, and the technology itself is intrusive. Some crude surveys have shown that people would travel as much as 50 miles rather than conduct the same conference over video facilities.

My own experiences with videotelephony have not been particularly encouraging—something seems to be missing that has not been electronically communicated. A British minister once told me that it would never work. To my embarrassment, he embraced me, saying, "I need to smell the person I'm dealing with." After the embarrassment finally passed, I understood what he meant. He meant "smell" as a metaphor for that human quality that the electronics had failed to convey. In my dealings with picturephone I also came to feel that it took too much of me, whereas the voice telephone demanded only a portion of my attention. Psychologists continue to study the efficacy of videoconferencing, but again the pace of learning about the ever-complex human interface does not match the pace of mindless technology.

Although the near future does not seem to promise the replacement of face-to-face meetings with electronic togetherness, new kinds of computer-mediated interactions among people will nevertheless be facilitated. ISDN in particular brings together the computer and voice-telephone environments. In multi-media conferencing participants can share an audio and a visual space. The audio space is the conventional speakerphone environment. The visual space is a computer-generated one of images and computer applications. In addition to the shared visual space, where everyone's computer-based images can be seen by everyone else, each person must have some individual method of pointing at objects appearing in the joint presentation.

As an example of multi-media conferencing, suppose that I call you and another friend. I may place the call directly from my computer workstation. When you answer the ring through your personal workstation, your computer screen becomes the shared visual space. Perhaps as a mnemonic reminder of the participants, small pictures of our faces appear in the upper right corner of the screen. These may well be still pictures transmitted during call setup from a "face server" computer file which has digitized images of everyone in the local telephone book.

In the course of the conference I wish to discuss with you a memo I have been writing. I call it up from a document data-base, and it appears on all our computer screens. As we talk about a section of the memo where you have some difficulty about the wording, you point to the sentence in question using a little pointer on the screen which is color coded with the same color that borders your picture in the upper part of the screen. Perhaps I edit the memo as we talk, and we agree on the new wording that is seen on the screen. I save the memo back to the document store.

Before we leave the call you tell me about a new idea you have. In order to describe your new widget, you call up a black-board program from your computer and begin to draw the widget. Again, all of us in the teleconference see your black-board. Your friend, who has been relatively silent in the con-ference, thinks your widget should be a little different. He draws his amendment on the blackboard. It is a shared space, so any of us can draw on it, just as if we were all standing to-gether around a physical blackboard. Now perhaps I am not so sure that your widget does what it is supposed to do. In order to prove my point, I ask my computer for a calculator program. As with any other application program, I can share the output among the participants, so that you may see my calculations as I progress. While this is going on, you decide to have a side conversation with me, since you believe that your friend's idea is not a good one and you want to ask my advice without offending him. That too is possible; the computer system mediates the conference so that side and sub conferences are easily arranged.

## The Further Future

In addition to ISDN there are other trends that should come to fulfillment in the near future. It seems certain that we will find some way to build and maintain a national data network with universal data connectivity for electronic mail and data transfer at a variety of rates. Perhaps this network will be built upon ISDN, and perhaps not. On the positive side of this question is the standardization and agreement on ISDN, and its all-digital, end-to-end connectivity. On the negative side are questions about how fast ISDN will spread, and the fact that it is defined only at the lower levels of the protocol hierarchy. In itself it does not define a usable network for computer communications, although one could be built upon it.

Computer networks will also become faster. There is already much work being done on defining new international standards for wideband ISDN (at about 150 million bits per second), and for the optical evolution of local-area networks (called FDDI, for Fiber Distributed Data Interchange), at 100 million bits per second. The National Science Foundation has led a movement toward even higher rates, funding a study of a billion-bit-per-second national network to interconnect research facilities. This is approximately the rate of the data transmission that takes place over the optic nerves that connect our eyes to our brain. What computers could do with such a rate remains to be seen, but it offers a grand challenge for some kind of human/computer breakthrough.

The evolution of cellular telephony should bring telephone access to the point of complete mobility. There are cellular advocates who believe that the diminishing cell size will eventually obviate fixed wiring in homes and offices. Whether or not this happens, there is no doubt that at some time in the not-too-distant future we will seldom, if ever, be out of touch with the possibility of access to communications. The term "personal communications" is used to describe this movement toward total freedom of access. With such access, of course, goes the burden of accessibility. To what degree do we want to be accessible to other people? At least we will have the choice. Furthermore, we should at last be able to call a person directly, rather than calling a place as we do today.

Standards and systems are being designed to give each of us a personal telephone number that knows where we are as we move. When you have lost touch with an old friend, it will only be necessary to dial his old telephone number, or perhaps his Social Security number; the network will know where he is. It seems, however, that there is a direct tradeoff between this accessibility and our cherished privacy. I love the idea of a Dick Tracy wristphone from which I can call anyone at any time; I just want to be the only person in the world to have one.

Not very far into the future we can see a fork in the road for communications. Right before the fork is a road sign that says HDTV (high-definition television). The sign points toward one of the roads ahead, but we cannot quite see from here which one it is. By now most people agree that HDTV will come; the only questions are who will bring it and over which facilities. The potential market is enormous, so all the main players are gathered for the match.

The local telephone companies hope that HDTV will come into the home through their optical fiber, thus creating a compelling reason for the wide-band pipe into the home and vindicating broad-band ISDN at 150 megabits per second.

The CATV companies also hope to be the conduit for HDTV. They too are upgrading to optical fiber trunking systems, although the last thousand feet into the home remain the province of coaxial cable for the time being. The companies who own satellite systems plan to lease their transponders for direct broadcasting of HDTV to small rooftop antennas. Although they could carry only one or two such channels, their satellites would offer a quick start to the industry. They are helped by the decreasing size and expense of the necessary home satellite dishes. Broadcasters, too, may offer HDTV over the air, but there is some argument about the vulnerability of the high-definition signal to the inevitable imperfections of broadcast reception—particularly multipath reception ("ghosts"). The ruling of the FCC that the HDTV signaling scheme must be compatible with existing television has given some impetus to conventional broadcast and CATV distribution, but the main game remains yet to be played.

The entertainment industry is also gearing up for HDTV; they see a new distribution channel for their product. They are possessive of their intellectual property, and they hope to find a new way to market it selectively without unduly diluting its value. The best solution for them would be the control offered by a switched fiber system that could feature pay-per-view programs. Although the satellite distribution system could also be turned on selectively at each home, the signal might be vulnerable to unauthorized interception and duplication.

There is yet another player in the HDTV stakes, and that is the system of video rental stores. Who would have thought a decade ago that the country would be inundated by neighborhood video rental stores? They too offer pay per view programs, and they could do so for HDTV also. Even though electronic distribution of first-run films would be more convenient, the video stores have become a social institution, and people might still prefer that means of acquiring entertainment. The video stores also have the potential of increasing their own convenience in competition with pay-per-view broadcasting. For example, they might offer delivery and collection services. The economics of physical delivery versus electronic delivery are yet to be established.

Whatever the outcome of the HDTV sweepstakes, the result is sure to shape the future of communications. The enormous bandwidth required by HDTV will be the making of whatever delivery vehicle is chosen. If that vehicle is fiber to the home, then perhaps all the other communications services can also ride that fiber, hitchhiking along with HDTV. If it comes by satellite or through video rental stores, then it may be hard to wire the nation with fiber to the home just for telephony. The game remains to be played; it is sure to be interesting.

### The Ultimate Goals of Communications

As we like at the more distant future, it is really impossible to conceive of what communications might offer. However, let us at least ask what the goals of communications should be. If we had complete mastery of technology, what would we ask from our electronic genie? Not so long ago I would have said that the ultimate aim of communications was to enable us to be somewhere else, so that we might do there what we do here.

Perhaps the culmination of that goal is what is known as *telepresence*.

It seems that we have always aspired to telepresence. In the Ice Age men drew pictures on the walls of their caves, depicting the animals that they hunted. Painting flourished during the Renaissance, but for realism it was supplanted by photography. Then came movies, sound, and color. Television brought a remote presence into our homes, and now we work toward high-definition television. We will probably have wall-size television screens in our homes. Someday we will undoubtedly invent a three-dimensional television that we like, perhaps using holographic image reconstruction. Perhaps, too, we will be able to choose our own viewpoint in the presentation, and move around as we wish.

Ultimately, we might communicate and reproduce all of the sensory input—sight, sound, touch, and smell—from a remote location. We might use robotics to communicate our own movements, so that we might explore the other place as we wish, turning our head this way or that so as to see in any direction. Our presence would then be, in every observable way, located remotely. However, whenever I think about using telepresence to visit a beach in Tahiti during a cold day in New York, I think about seeing other telepresence robots walking toward me on the beach. There still is something to be said for interacting with real people. As much as I might wish to be represented at a conference by my telepresence robot, I would not want to deal with others in this fashion. How demeaning!

Although some years ago I thought of telepresence as the ultimate goal of communications, now I have become still more greedy. It is not enough to be where we are not. We should also be able to be *when* we are not. By this I do not mean time travel, which I will wistfully rule out, but a flexibility in time displacement for communications. We already see this happening in video tape recorders, telephone answering machines, and the like. We simply do not have the freedom to schedule interactions to everyone's convenience; thus we must find a way to achieve the purposes of communications when we are unavailable.

Today flexibility in time is accomplished best by a human intermediary, usually a secretary. A secretary can act as a communications surrogate, or can interactively seek information. An answering machine can only accept messages. Is the gap between the two uncrossable, or can machines extend their help in the direction of increased intelligence and autonomy? One notion is that of an agent, an intelligent machine or process that acts on behalf of its owner in some specified matter. In the acquisition of information, for example, an agent (which has been called a "knowbot") could "march" out into the network and stir around until it found what it was instructed to retrieve. A simpler example would be setting up a meeting. The agent would check the calendars of proposed participants and find an appropriate date and place, and then verify with all concerned. This is quite a bit harder than it sounds, however, because of the sensitivity most people have with their own calendar information and calendar control. But since secretaries can do this successfully, it offers an existence proof that there must be an algorithm that can be followed, given sufficient information and intelligence. Many other common tasks are similarly difficult, but demonstrably achieveable by a surrogate.

The future will not be determined by technology alone. Not by a long shot. In recent years engineers have had dreams of new communications services—videotelephone, information and education in the home, energy management, appliance control, security, and so forth. These have all been technologically feasible and economically viable, but they have been rejected by society. There has also been brave talk about displacing travel and working at home through telecommunications. Again, the social factors were ignored in these projections, and the actual displacements have thus far been small.

Progress on the front of human communications has been much slower than the advance of technology. How do we really communicate with one another? What happens in face-to-face meetings that we need to emulate at a distance with technology? Unfortunately these are almost unanswerable questions. Human beings are complicated and finicky. We will be hard to satisfy even with future technology. On the other hand, machines will be built around the communications technology as it evolves. There is no such thing as face-to-face for a machine. Machines everywhere will be bridged together to form a pool of intelligence and power. In the end, of course, it matters only that the power that emerges works to the benefit of mankind. If experience is any guide, more communication is better. The more things are open, the more we are interconnected, the better off we are. This is the promise of future communications.

# Public Policy for the Information Age

*John L. Pickitt*

The health of the worldwide information-technology industry influences, directly or indirectly, almost every public-policy concern of computer and business equipment manufacturers in the United States. Moreover, these companies are shaped by international events, since they market products globally, source componentry abroad, ship materials and products across borders, are both vendors to and customers of corporations based in other countries, and are partners in joint ventures with foreign-based corporations.

While US manufacturers have an obvious, driving, and ongoing interest in Washington's policies, they also must focus on policies affecting corporate life in Canada, Japan, the European Community, the Pacific Rim, and other nations and areas where these manufacturers operate. US companies will also increasingly address conditions and market opportunities in Eastern Europe. These concerns are compounded by multilateral issues: How can multilateral treaties governing trade be improved? What is the role of bilateral agreements in facilitating trade?

Among the major public-policy issues that will shape computer and business equipment manufacturers' engagement in international trade in the 1990s are the maturing market at home, the soaring cost of high-technology innovation and production, the growing role of newly industrialized countries in trade negotiations, the web of challenges as the European Community increasingly becomes an integrated market, and the prospects of shrinking demands for defense-related technology.

Domestically, US policies affecting a company's ability to compete at home and abroad will be high on the list. Internationally, any agreements, standards, or rules that affect trade will be of keen interest.

At the beginning of the 1990s, information-technology companies generated 9 percent of the United States' gross national product. Between 1960 and 1988, employment in information-technology production increased by 1,240,100 to a total of 2.4 million workers. This is an average annual increase of 3 percent. During the same period, employment in manufacturing grew a mere 0.5 percent per year. Clearly, the manufacturing and maintenance of computers, printers, facsimile machines, copiers, and related products has embodied steady and growing opportunities for millions of Americans.

Moreover, the presence of information-processing technology has reshaped jobs in medicine, engineering, graphic design, education, and other areas, and it is projected that by the year 2001 every office worker in America will use a computer on the job. High-tech innovation has also led to the creation of jobs directly tied to the technology: systems analysts, programmers, computer and peripheral equipment operators, keypunch equipment operators, and computer service technicians have a place in every industry. By 2001, the number of such jobs should reach 3 million.

Wages have been noticeably on the rise in the industry. In 1988 dollars, the average hourly wage paid to the workers in computer and business equipment manufacturing went from $3.00 in 1965 to $4.82 in 1975 and $9.32 in 1985, and it has continued to go up steadily since then. In 1975, when computer and data processing jobs became more prominent in the mix of opportunities in the United States, average hourly wages for the computer and DP service workers were $5.26. Ten years later, they were up to $10.98, they had reached $13.11. An astonishing rise has also taken place in the telecommunications equipment and services industry sectors. Workers in the equipment sector received $2.86 an hour in 1965 and $11.62 by 1988; those engaged in telecommunications services started at $2.70 in 1965 and were up to $13.45 in 1988.

Continuing this momentum in US job and educational opportunities hinges on US computer and business equipment companies' worldwide technological competitiveness. Yet foreign competitors often have better manufacturing environments than the United States. For example, Japan's 1988 investment in plant and equipment ($498 billion) surpassed that of the United States ($487 billion). In 1989, Japan's spending reached about $560 billion, while the estimated US investment was just $514 billion. Consider also that in 1970 the US was spending 2.5 percent of its gross national product on research and development. Twenty years later, it is spending about the same. In the same period, Japan's allocation started at 1.8 percent of GNP and is now almost at 3 percent of a much more swiftly expanding GNP.

Foreign competitors often receive significant incentives from their governments supporting growth in research and development programs and facilities. The Japanese industry, for example, had the advantage of a permanent 20 percent research and development tax in addition to a 7 percent credit on the purchase of equipment used for basic research for several years. In contrast, the United States does not have any permanent credits to provide incentives for domestic companies to conduct R&D. A shaky system of temporary refund measures—tax provisions with built-in expiration dates—is what US companies face. The wider concern—shared by individual citizens as well as corporations—is the degree to which the US risks being unprepared to participate in the highly complex world of corporate R&D.

### National Excellence in Education

The United States must make excellence in education a national priority so that new employees are at least as well prepared to contribute to the economy as are citizens in other countries. Though it is true that the aggregate number of US students earning computer science degrees has increased over the past twenty years, that achievement is only one block in a large structure. Among the gaps in the structure are the insufficient number of qualified candidates for scientific and technical jobs and the fact that some key trading partners are outpacing the US with their educational capabilities.

At the end of the 1980s, the Council on Research and Technology (CORETECH) produced a report entitled "Meeting the Needs of a Growing Economy: The CORETECH Agenda for the Scientific and Technical Workforce." CORETECH's recommendation of a program aimed at upgrading US education from the elementary level through the continuing-education level reflects the thinking of its 162 major corporate, university, research, and association members, including the Computer and Business Equipment Manufacturers Association (CBEMA). The report is an urgent call for policy initiatives that involve US corporations, universities, and the federal government. In summary, it recommends the following:

• Corporations should work with math and science teachers to expose them to actual research settings, should bring students into contact with scientists and researchers by providing them with part-time jobs and internships, and should create a work atmosphere that encourages employees to pursue further training and education.

• Universities should create opportunities for students to meet and work with university researchers, should offer continuing-education programs for math and science teachers, should hire female and minority faculty members, should encourage reentering students (often women) to pursue math, science, and engineering degrees, and should provide convenient continuing-education opportunities that will reach the more dispersed, non-urban workforce.

• The federal government should support programs to upgrade the skills of math and science teachers, should provide incentives for talented graduates to become teachers, should assist institutions with the modernization of academic research facilities, should help elementary and secondary schools with large numbers of minority students to improve science and math education, and should expand or establish in each research agency fellowships and internships for female and minority students. Further, the government should reverse the loan/grant imbalance. (The erosion of grants in the face of the steadily increasing costs of education have forced students to rely heavily on loans.) Tax incentives for industries to offer continuing-education opportunities for employees and to support the development of high-quality two-year technical programs are also needed.

• Finally, the US must foster communications and collaborations among government, industry, and academic partners to maintain the excellence of the scientific and technical workforce. Corporations and universities should call for and participate in forums that bring together national, state, and local officials with educators and corporate executives. The federal government should prepare a national assessment of current and projected workforce training and retraining needs. And the government should convene a national congress of state-level forums and institutions that are working to upgrade the skills of their scientific and technical workforce. The congress should result in models for the states to use, as well as proposals for federal participation.

### Research, Development, and Taxes

Given appropriate tax incentives to engage in research and development in the United States, US-based companies will contribute even more dramatically to America's quality of life and balance of trade. Two measures are needed to boost those incentives. One is designed to encourage continuing annual increases in corporate R&D spending; the other would revise rules in the tax code which actually give some companies reasons to move R&D facilities outside the United States.

The R&D Tax Credit, first adopted in 1981 as part of the Economic Recovery Tax Act, has been the only provision of the US tax code that aims to increase applied industrial research. It provides for a 25 percent credit on the annual increase in a company's R&D spending above the company's average R&D spending for the prior three-year period. This original credit expired on December 31, 1985. In 1986, the credit was extended through 1988, but the rate was reduced from 25 percent to 20 percent. Later, in their zeal to try to reduce the US deficit, key members of Congress—ignoring the long-term implications—called the credit a "revenue loser" and would not support its extension, even though such action undermines support for R&D and seriously erodes the competitiveness of US products.

Japanese companies—generally considered to be the major competitors to US high technology—have a more supportive set of incentives for research and development. Their tax policies acknowledge that R&D undertakings generally are multi-year projects linked to considerable risks for an individual company, and taxes are therefore structured to encourage continuing growth of industry investment in R&D.

Innovation may be the most obvious result of enhanced R&D, but there are corollary results that affect a company's competitiveness and the nation's economy. A study conducted for CORETECH in the late 1980s by Martin Baily and Robert Lawrence of the Brookings Institution found that a permanent R&D tax credit—structured as it was originally, with a 25 percent rate, and claimable only for increases in R&D spending—could boost the GNP by up to $17 billion annually beginning in 1991. Surprisingly, the tax code of the early 1990s actually contains disincentives (section 1.861.8) to US companies to build and maintain their R&D facilities inside the United States. By restricting the deduction of R&D expenses from US income, Internal Revenue Service rules seriously disadvantage companies that do most of their R&D in the US but sell products worldwide.

Allowing domestic R&D expenses to be deducted from domestic income, without building in constraints that disadvantage companies with operations in different nations, would be an incentive for American companies to keep R&D at home. But so far, the IRS rules force companies to allocate a certain amount of the domestic R&D expenses to income earned in other countries. The effect is that, by financially penalizing companies that do R&D in the United States and then export overseas, Section 861 encourages them to move R&D facilities and the associated jobs to other countries where such expenses could be deducted. Although there are many business reasons for selecting an R&D site, the US tax code should not encourage US companies to take their R&D to other countries.

### Open Trade Policy

Policies concerning how the US government handles difficulties with trading partners should also be long-range; the antithetical short-term view is protectionism. "Unfair trading policies," such as dumping (selling products for less than their fair market value), piracy (duplicating and marketing, without permission, intellectual property which is protected by copyright, patent, or trademark, and market-access barriers undercut US high-technology companies in world markets. US trade law provides numerous tools for seeking remedies for unfair trading practices on a bilateral basis. A careful use of those tools has yielded positive—although perhaps not yet satisfactory—results. The Brazil Informatics 301 case is an example. At the same time, there are instances where the outcomes are still disputed; the semiconductor arrangement perhaps best fits into this category.

These controversies beg the long-term question: Will this action improve access to a closed foreign market (or stop dumping, or eliminate piracy)? The corollary, of course, is Will this action improve conditions for only one company or industry sector, while it damages others?

US trade law in the early 1990s also provides a means for opening foreign markets and addressing unfair trade practices on a multilateral basis—that is, authority to negotiate in the Uruguay Round of the General Agreement on Tariffs and Trade (GATT). In view of the global nature of the information-technology industry, global solutions are most likely to produce positive results. The Uruguay Round provides an opportunity to identify and adopt those solutions. For example:

• The adoption of an intellectual property rights regime by the GATT will protect the investment of the industry in developing new technologies and enhance the industry's competitiveness.

• Amendments to the GATT anti-dumping code and the development of a multilaterally agreed-upon rule of origin will prevent the unfair use of national dumping laws and rules of origin to protect domestic industries from fair competition or to force investments.

• Extension of the GATT procurement code to cover more government agencies and new sectors will stimulate competition and the flow of trade for the information-technology industry.

• Broadening the scope of the GATT to include principles and sector-specific rules for international trade in services will work to expand trade and investment in services.

• GATT adoption of an agreement aimed at eliminating trade-restrictive and trade-distorting effects of government investment policies and practices will increase the ability of information-technology companies to make rational business decisions.

• Reducing if not eliminating tariffs on finished products and componentry will increase the competitiveness of the information industry.

These are the promises of GATT. Nevertheless, steps in the direction of opening markets and adopting uniform rules increase business certainty. Steps away from bilateral negotiations that heighten trade friction and require resource investments to avoid unwelcome outcomes—resources that could be used for more productive purposes—are steps in the right direction.

### Responsive Export-Control Policy

Though all US industries could be directly affected by trade sanctions, quotas, and many other elements of trade policy, the US high-technology sector has been particularly affected. For a generation, exceptionally restrictive controls have been imposed on exports that could bolster the military strength of the Warsaw Pact. Yet the proliferation of the capability in many nations to produce and market high-technology products is reason in itself to argue for reform of the US export-control system. Political events that undercut Soviet military capability offer additional arguments for reforming the system.

During the rest of the 1990s there need to be thoughtful adjustments to US export-control policies that are consistent with the worldwide advances in technology and consistent with the level of military threat. Such adjustments should also be consistent with our need to reduce the US trade deficit, to balance our budget, to enhance US industry's international competitiveness, and to retain US preeminence in high technology.

For example, the Export Administration Act should be amended to recognize that a multilateral export-control system that fully supports US national interests must both deny militarily significant technology to potential antagonists and simultaneously enable US industry to compete equally with companies in allied nations in the international market.

### Coordinated International Standards

One element in the mix of critical public-policy issues does not necessarily involve government. This element is industry standards. In the United States these standards emerge from mixed groups of volunteers (users, manufacturers, and other interested parties), not solely from officials.

How the international harmonization of information-technology industry standards will facilitate trade (or, conversely, how the lack of harmonization will profoundly inhibit trade) is one facet of the issue. Whether or not foreign governments will try to make US companies jump through fiery hoops to prove that their products conform to certain standards, or use standards as nontariff trade barriers, is another. Users' demands that one company's innovation be compatible with competing brands, that a disk written by one computer be easily read by another, and so on, are tied to standards. These demands will only get more specific as information technology evolves.

The voluntary standards system in the US, coordinated by the American National Standards Institute, brings together manufacturers, users, and other interested parties to agree on details of a programming language or on how two pieces of hardware will interconnect. America's trade partners pursue similar efforts. International standards bodies have, in the past, served mainly as forums for harmonization of the standards accepted by different countries; in the information-technology industry, a significant percentage of standards are initiated at the international level.

Imagine the enormity of a trade barrier between, say, the United States and the European Community if divergent technical standards are developed. The EC is establishing EC-wide standards with associated testing and certification so that products meeting "essential" health, safety, and environmental requirements (e.g., Euro-beef, Euro-beer, Euro-bread, Euro-ice cream) as stipulated by new EC directives can be marketed freely throughout the community. These new directives could have a powerful influence on international standards and a major impact on current and future access of US producers to the EC market.

EC Technical Harmonization Directives or the programs implemented by them could be framed, even inadvertently, to exclude or hinder the entry of US products into the European market. Of further significance could be the effect on US imports from Europe; European products are likely to be more competitively priced, owing to economic benefits derived from standardization. In spirit there is US corporate support for the EC's unified approach to standards as a way to liberalize the movements of goods throughout the community. However, there is also concern that the EC standards process could obviously create new barriers between the EC and its trading partners. Voluntary international standards therefore need to be EC-wide standards.

Intertwined with the setting of standards is certification, the act of documenting that products conform with accepted standards. In the area of certification, the major issue is access to all testing and certification programs in the EC (or elsewhere) for goods produced in the United States. This access should be on the same basis as for goods produced in the EC, or in another country or region to which US companies export. Certification, like standards, can be used as a nontariff trade barrier, especially when the certification authority is a national or regional government. A government can require that conformance be tested in its country or region, by its people, in its testing houses, using its tests (which may be unique or unreasonably stringent) and test tools (which may be different or unique). The US information industry's long-held view is that manufacturers' self-testing and declaration of conformance provides the best method for assurance that their products meet the standards. They have primary responsibility and accountability for implementation of the standards in their products.

Why conform? Manufacturers know they can't sell what people don't want, and, once the standards for a product have been developed they probably reflect what most people desire in the product. This is the basic argument against having regional, national, or international certification "schemes" or regulations to ensure proper implementation of information standards.

At the heart of the standards discussion is the fact that manufacturers must answer users' needs and users must clearly define these needs. Productive standards development and conformance testing will depend on

developing voluntary standards through an open consensus process,

harmonizing national and international voluntary standards,

the adoption of voluntary consensus standards whenever feasible and consistent with law and regulation, in preference to government-developed standards,

engaging in manufacturer self-testing and declaration of conformance—the best method for ensuring that products conform to standards,

supporting third-party conformance testing only for those who do not have the capabilities or, for other reasons, do not choose to self-test, and limiting certification criteria to matters of health, safety, environmental protection, national security, and (where appropriate) harm to telecommunication networks.

### Conclusion

Inventing the future of computing and communications requires a long-term approach with a global perspective on trade. Specifically, it must involve pro-R&D policies that at least provide US companies with an operational environment as favorable as that of their international competitors; a national policy on excellence in education so that the workforce is prepared to meet the job challenges posed by technological innovations and international competition; laws and regulations that support open trading; export-control policies that reflect changes in the marketplace and the international climate, so that resources no longer needed for military applications can be used to bolster international trade, to reduce the trade deficit, and to build the economic foundation to ensure national security if tensions resume; and, finally, international harmonization of industry standards, and a reasonable approach to conformance testing once standards are accepted. (Standards already have an almost immeasurable impact on the shape and strength of the industry, and their potential to either facilitate or block international trade will grow.) In sum, the interaction of technology and public policy is shaping the environments and operations of the millennium.

# About the Authors

*Denos C. Gazis* is Assistant Director of the Semiconductor Science and Technology Department of the IBM Research Center in Yorktown Heights, New York. He served previously as Director of the General Sciences Department, as Technical Advisor to the IBM Vice President and Chief Scientist, as a Member of the Research Review Board, and as Assistant Director of the Computer Sciences Department. He has been a Visiting Professor at Yale University. Dr. Gazis has also served as a consultant to the Institute for Defense Analysis and to the Congressional Office of Technology Assessment. His books concern traffic science and the theory of vibrations of solids. Dr. Gazis has his first degree in Civil Engineering from the Polytechnic in Athens. While a Fulbright Scholar at Stanford University, he received an MS in Engineering Mechanics. His PhD in Engineering Sciences is from Columbia.

*Kathleen Bernard* is the Director of Science Policy and Technology at Cray Research. She recently served as an American Electronics Association Fellow in the White House Office of Science and Technology Policy. She currently serves as a member of the Technology Subcommittee of the NASA Advisory Committee. She holds an MBA and has a BS in mathematics from the University of South Carolina.

*Patrick Gelsinger* is an engineering manager in Intel Corporation's Microcomputer Division. He heads the 80486 design team, having been a member of the 80386 design team. He has ten publications on VLSI design, VLSI CAD, VLSI testing, and on computer architecture, in addition to one patent and a co-authored book, *Programming the 80386* (1987). He has a BS

379

from Santa Clara University and an MS from Stanford in electrical engineering.

*Paolo Gargini* is a Technology Coordinator at Intel. In 1980, he became the manager of Basic Technology Development for microprocessors and SRAMs, becoming responsible for developing the building blocks of HMOS III and CHMOS III—technologies which were used throughout the 1980s. As Intel continued to develop parallel technologies at multiple sites, Dr. Gargini became the One-micron Technology Coordinator. Five years after receiving his doctorate in Electrical Engineering from the University of Bologna, he received a doctorate in physics from the same institution. He has done research at LAMEL in Bologna and at the Stanford Electronics Laboratory. He holds patents on EPROM and Interconnection technologies.

*Gerhard Parker* is a vice president of Intel and Director of Component Technology and Development. He is responsible for the continuing development of the silicon processes by which most of Intel's microelectronic products are made, and for design groups which lay out the final circuitry. He received his BS, MS, and PhD in Electrical Engineering from the California Institute of Technology.

*Albert Y. C. Yu* is a vice president of Intel and General Manager of the Component Technology and Development Group. He is responsible for the development of all VLSI logic, ASIC, and memory products. His BS in Electrical Engineering is from the California Institute of Technology, and his MS and PhD in electrical engineering are from Stanford.

*Harry Tennant* is a Texas Instruments Fellow and Chief Technologist of the Information Technology Group of Texas Instruments, Inc. Dr. Tennant invented the concept of menu-based natural-language understanding systems. He hosted and was responsible for the content of Texas Instrument's Artificial Intelligence Satellite Symposia, a series of video broadcasts on applications of artificial intelligence. He is author of the book *Natural Language Processing: An Introduction to an Emerging Technology* (1981), and was selected by *Science Digest* as one of America's 100 outstanding scientists under age 40. His BS, MS, and

PhD are in Information Engineering and Computer Science from the University of Illinois.

*George H. Heilmeier* is Senior Vice President and Chief Technical Officer of Texas Instruments. In 1968, he received international recognition for his discovery of several new electro-optic effects in liquid crystals, making possible for the first time the electronic control of the reflection of light. While a White House Fellow during 1970–71, he served as a Special Assistant to the Secretary of Defense. He was then appointed Assistant Director for Defense Research and Engineering and, in 1975, Director of the Defense Advanced Research Projects Agency, where he initiated major efforts in space-based lasers, stealth aircraft, space-based infrared technology, and artificial intelligence applications. Dr. Heilmeier holds 15 US patents and is a member of the National Academy of Engineering and of the Defense Science Board. His BS in Electrical Engineering is from the University of Pennsylvania, and his MSE, MA, and PhD in solid-state materials and electronics are from Princeton.

*William R. Johnson, Jr.* is Vice President, Telecommunication and Networks, for Digital Equipment Corporation. He is responsible for the engineering and marketing of DEC's networks and communications products. He is also responsible for the engineering of DEC's image processing and artificial intelligence products, and he has functional responsibility for international engineering. His BS in Electrical Engineering is from Penn State, and his MSEE and MBA are from Northeastern University.

*Al McBride* is the director of future technology for Tandem Computers. His responsibilities include consulting on new technology ventures in which Tandem invests and defining new product opportunities for Tandem. He joined Tandem in 1980 as director of VLSI technology. He had previously been at IBM for 16 years, where he was involved in CPU design, CAD systems, and the development of one of the industry's first NMOS microcomputer chips. Mr. McBride holds BSEE and MSEE degrees from Berkeley.

*Scott Brown* has been a strategy consultant for Tandem Computers for two of his eight years at Tandem, where he previously managed software development. He had previously been a development manager for Burroughs, responsible for designing network architecture. He has a BS in Computer Science from Utah State University and an MBA from Santa Clara University.

*Roger E. Levien* is Vice President, Strategy Office, Xerox Corporation. He helps establish the company's strategic direction while monitoring worldwide market and technology trends as well as managing the strategic planning process. Before joining Xerox, he was director of the International Institute for Applied Systems Analysis (IIASA) in Laxenburg, Austria, from 1975–81, and was head of the System Sciences Department and Deputy Vice President of the Rand Corporation from 1960–74. He has taught at UCLA. His books are *The Emerging Technology* and *R&D Management*. Dr. Levien's BS in engineering is from Swarthmore, and he holds an MA and a PhD in applied mathematics from Harvard University.

*Olaf Olafsson* is Vice President of the Sony Computer Peripheral Products Company. He is responsible for overall management of the Optical Products Group and also coordinates Sony's US multi-media activities. A physicist from Brandeis University, he has been with Sony since 1986. Mr. Olafsson is the author of two best-selling novels published in Scandinavia in 1986 and 1988.

*Harold F. Langworthy* is Director, Imaging Information Research Laboratories, Commercial & Information Systems Group, of Eastman Kodak, having previously directed the company's Physics Division. He has a BS in mathematics from Rensselear Polytechnic Institute and a PhD in mathematics from the University of Minnesota.

*Timothy Dickinson* is a strategic planner and economist with Future Technology, Inc. He also consults for banks and international trading firms. Mr. Dickinson teaches occasional courses in military and economic strategies at the US Defense Department and Magdalen College, served for several years as the Washington editor of *Harper's* magazine. He has a BA from Oxford.

*Lee W. Hoevel* is chief architect for NCR Corporation's Workstation Products Division. He previously held positions as director of NCR's Advanced Human Interface Center and director of Systems Architecture in the Research and Development Division. He has also worked at IBM's T. J. Watson Research Center as Senior Manager of Environmental Architectures. Dr. Hoevel holds degrees in Economics and Mathematics from Rice University, pursued graduate studies in Computer Science at Stanford, and has a PhD in Electrical Engineering from Johns Hopkins.

*Robert W. Lucky* is Executive Director, Research, Communications Sciences Division, of AT&T Bell Laboratories. His division researches methods and technologies for future communication and computing, including current emphasis on lightwave systems, multiprocessor computer systems, robotics, artificial intelligence, and new physical devices for optics and electronics. His early work included the invention of the adaptive equalizer—a technique for correcting distortion in telephone signals which is used in all high-speed data transmission today. Dr. Lucky is the author of the recent book *Silicon Dreams*. He has served as Vice President and Executive Vice President of the Institute of Electrical and Electronics Engineers, and has been editor of three different technical journals, including the *Proceedings of the IEEE*, in addition to continuing his column in IEEE's *Spectrum* magazine. He is a Fellow of the IEEE and a member of the National Academy of Engineering. He is currently Chairman of the Scientific Advisory Board of the US Air Force. He holds BS, MS, and PhD degrees, as well as an honorary doctorate of engineering, from Purdue University.

*John L. Pickitt* is president of the Computer and Business Equipment Manufacturers Association (CBEMA), which represents the leading companies providing computer, business, and telecommunications hardware, software, and services. As a Lieutenant General in the US Air Force, he served as Director of the Defense Nuclear Agency, analyzing computers and communications systems in defense environments, including space. Before his years at the Pentagon he held multiple command responsibilities, including Deputy Commander in Chief of UN Forces in Korea and, in the US, Commander of the 1st Air Force. He has special expertise in applying high technology to national defense programs, and he designed and tested the first rocket engine test facility for the US Air Force Institute of Technology. He is a graduate of the US Military Academy at West Point as well as of the US Air Force Command and Staff College. He has a MS in Aeronautical Engineering from the US Air Force Institute of Technology.

*Derek Leebaert* is a professor of management at Georgetown University's Graduate School of Business. He served as Chief Economist of the Computer and Business Equipment Manufacturers Association (CBEMA) from 1985 to 1988, and continues as a consulting economist. He has taught program analysis in the Senior Officers' course at the Pentagon and still works on national security policy, and on technology issues and applications, within the US defense community. While a postdoctoral fellow and visiting scholar at Harvard University's Center for Science and International Affairs, and at the School of Government, from 1975–84, he was a founding editor of *International Security* and of the *Journal of Policy Analysis and Management.* Dr. Leebaert was thereafter a founding editor of the bi-monthly magazine *The International Economy.* He has a D.Phil. in economics from Oxford University and an MPP from Columbia University, and he was a Ford Fellow at MIT.

*Arthur C. Clarke, CBE,* who contributed the forecord, is the preeminent writer of science fiction. His many books include *2001: A Space Odyssey* and *2010: Odyssey Two.* He has also written the nonfiction *Profiles of the Future,* which former Citibank Corp. chairman Walter Wriston called the classic work for business strategy. Mr. Clarke co-broadcast the Apollo 11, 12, and 15 missions with Walter Cronkite and shared an Oscar nomination for the film version of *2001.* He also introduced the original notion of communication satellite. He is a Fellow of the University of London and serves as Chancellor of the University of Moratuwa, where he resides in the island country of Sri Lanka.

# Index